TRAVEL AND GEOGRAPHY IN THE ROMAN EMPIRE

TRAVEL AND GEOGRAPHY IN THE ROMAN EMPIRE

Edited by Colin Adams and Ray Laurence

London and New York

First published 2001
by Routledge
11 New Fetter Lane, London EC4P 4EE

Simultaneously published in the USA and Canada
by Routledge
29 West 35th Street, New York, NY 10001

Routledge is an imprint of the Taylor & Francis Group

Typeset in Garamond by Keystroke, Jacaranda Lodge, Wolverhampton
Printed and bound in Great Britain by St Edmundsbury Press,
Bury St Edmunds, Suffolk

British Library Cataloguing in Publication Data
A catalogue record for this book is available from the British Library

Library of Congress Cataloging in Publication Data
Travel and geography in the Roman Empire / edited by Colin Adams
and Ray Laurence.
p. cm.
Includes bibliographical references and index.
1. Rome—Geography. 2. Rome—Historical geography.
3. Roads, Roman. 4. Travelers—Rome—History. I. Adams, Colin, 1967–
II. Laurence, Ray, 1963–
DG30 .T73 2001
913.704'6—dc21 2001016013

ISBN 0–415–23034–9

CONTENTS

FIGURES

CONTRIBUTORS

Colin Adams is British Academy Postdoctoral Research Fellow and Lecturer in Ancient History at the University of Leicester.

Kai Brodersen is Professor of Ancient History at the University of Mannheim and Visiting Research Fellow at the University of Newcastle.

Jon Coulston is Lecturer in Ancient History at the University of St Andrews.

Anne Kolb is Professor of Ancient History at the Historisches Seminar der Universität Zürich.

Ray Laurence is Lecturer in Ancient History at the University of Reading.

Benet Salway is Lecturer in Ancient History at University College London.

PREFACE

Five of the chapters included in this volume were originally given as papers at the Society for the Promotion of Roman Studies Roman Archaeology conference at Durham in April 1999. The editors would like to thank Martin Millet and Simon James for their assistance in organising the panel on transport.

The editors would like to thank the following for their advice: Richard Alston, Kai Brodersen, Jon Coulston, Thomas Harrison and Greg Woolf. The support and patient encouragement of Richard Stoneman and Catherine Bousfield at Routledge launched the project to create the volume and have eased some of the trials and dilemmas faced by the editors. Figure 3.2 was supplied by Lawrence Keppie. Figure 6.4 appears with the permission of Mike Bishop. Figure 7.1 was drawn by Debbie Miles-Williams. Figure 7.2 appears by permission from the Egypt Exploration Society. Figure 7.3 was enhanced by Lucy Farr. Figure 8.2 appears with permission of Hugh Davis. Colin Adams wishes to thank the British Academy for the award of a British Academy Postdoctoral Fellowship, which has afforded time to carry out this and other research projects. Ray Laurence wishes to thank his colleagues at the University of Reading for their support.

1

INTRODUCTION

Colin Adams

The purpose of this volume is to explore connected themes on geographical knowledge and travel in the Roman world. It is a growing field, built upon much existing scholarship, and new evidence is coming to light which might revolutionise our understanding of how the Romans viewed the world in which they lived and how they travelled in it. The recent discovery on papyrus of what might be the only ancient map from the Roman period will certainly add complexion to this.[1]

The chapters are founded upon a series of five papers delivered by Adams, Brodersen, Coulston, Kolb and Laurence at the 1999 Roman Archaeology Conference in Durham, to which Salway's paper has been added. The richness of the subject of transport and travel finds further reflection in the different approaches used by the contributors, and by the different forms of evidence, literary, epigraphic, papyrological, iconographic and archaeological, which form the basis of their papers.

Travel and communication are dynamics which were central to the Roman empire. Its sheer size and diversity demanded that there was an efficient system of communication in order for government to take place. This is not something which can only be seen with hindsight – Roman writers recognised the importance of communication, as we can see from a well-known passage from Aristeides of Smyrna, in which he recognised the Roman 'world-state':

> Could not every man go where he wished, without fear? Are not all harbours busy, are not mountains as safe as cities? Is there not charm in all fields, from whence dread has vanished? There are no streams impassable, no locked gulfs. The earth is no longer iron, but clad anew for a feast. Hellenes and barbarians may wander from their own homes to arrive at their own homes; the Cilician Gates, the narrow sandy roads to Egypt through Arabia present no terrors of mountain pass, torrents, or savages: to be the emperor's subject, to be a Roman is the one talisman. Homer had said, 'The earth is common to all'; it was now realised. You have measured the earth, bridged the rivers, and made roads through the mountains, and ennobled all things. The world

> need be no more described; no laws or customs retailed; for you have been the leaders for everyone, have opened every gate and given every man his freedom, to see all with his own eyes. You have conferred equal laws on all, and repealed conditions entertaining to the mind, and intolerable in reality; and merged all nations into one family.[2]

This speech was delivered in honour of the first Antonine emperor and is clearly thick with eulogy. However, it illustrates the importance of communication on a number of levels. There was a civilising force – the idea that Roman citizenship was a unifying state. But more than this, the Romans had provided the means for the world to be common to all.[3] Roman imperialism, displayed in the control of populations and the Romanisation of peoples, was also imposed upon the earth itself.[4] Land was measured and divided, rivers bridged, canals dug joining rivers and seas, and roads were cut through inhospitable country.[5] It became easier to travel – the long journeys made by envoys during the Hellenistic period, which were so fraught with danger and uncertainty, now became much easier and safer. Roman authors boasted of the *pax Romana* and the eradication of piracy and brigandage throughout the empire. This encouraged trade and transport, and Rome became the centre of this communication.

But was this new-found mobility something which affected all Rome's subjects; who were the travelling public? And how was communication organised? How did people know how to get from A to B? Did maps exist, or did itineraries serve as route-finders? Was it possible to find one's way, for example, from Rome to Brundisium, or from Antioch to Ephesus, using an itinerary in rather the same way as we might use a London Underground map or simplified plan for tourists that highlights the ease of travel to a place. Brodersen and Salway discuss these questions in detail, developing the fundamental work of Dilke.[6]

Brodersen surveys the existing evidence for ancient maps, and concludes that there is not enough to be certain that Romans had scale maps, or indeed needed them. Itineraries provided the information necessary for travel, and he compares this to modern road maps, which merely provide information on routes and junctions, but still permit efficient travel. *Itineraria adnotata* (annotated itineraries) provided enough information to travel along one route; *itineraria picta* (illustrated itineraries), such as the Peutinger Table or the recently discovered Artemidorus papyrus, permitted travel along any number of routes to several destinations. This interconnection of road systems, and indeed the fact that road, riverine and sea travel were all part of a complimentary system, is central to the work of Laurence on Roman Italy.[7] Salway argues that inscriptions should be seen as complementary to such *itineraria*, and with this in mind we should not ignore a letter of Pliny the Younger, who discusses the importance of roads and milestones in gaining directions to people's property.[8] In the case of the Peutinger Table, he suggests that placenames are important to our understanding of it as a tool for travel, and as such, it belongs fully to the itinerary

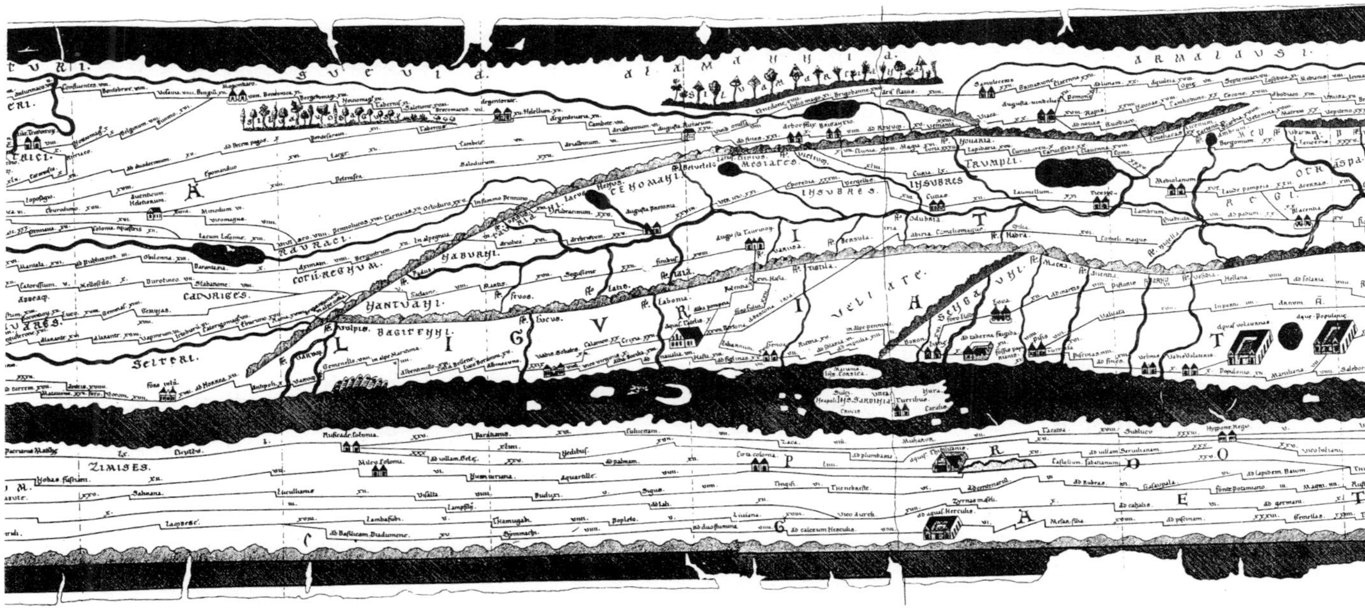

Figure 1.1 Peutinger Table: section showing southern Germany, northern Italy and Numidia (after K. Miller, 1887). (With permission from the Österreichische NationalBibliothek.)

tradition, rather than a distinct pictorial one. The Peutinger Table comes from a single *pictorial* tradition, but the information preserved on it derives from itineraries. These latter derive from disparate and different sources; not from government sources, such as the *cursus publicus*, but from more public data displayed on milestones and *tabellaria*, which he argues are lists of stages and distances, and from information derived from survey and road-building.

It is difficult to accept, however, that the government and *cursus publicus* did not have information concerning travel available to its officials. The communication infrastructure formed the backbone of provincial government. The imperial period saw a huge leap forward from the seemingly independent provincial governors of the republic, to a far more centralised system of government illustrated by Pliny the Younger's correspondence with Trajan. Such themes form the basis of the work by Kolb, and her chapter here considers the role of the *cursus publicus* or imperial post service. There are clear links between this and the development of itineraries which have been preserved. It is fair to say that the road network of the empire served the needs of the state, both administrative and military, before those of private individuals; but the latter could certainly benefit. Adams shows that considerable numbers of travelling officials and soldiers could have used *mansiones*, the staging posts of the *cursus publicus*, at any one time, and there is good reason to believe that there was a significant amount of communication around the empire. Given this, it is likely that officials had access to information to facilitate their journeys, but Adams suggests that this information existed for another significant purpose, the administration of funds available to officials to pay for their travel. In other words, as well as providing information for travellers, itineraries served an additional fiscal purpose.[9] Travel and communication permeated all levels of state bureaucracy, the fiscal sphere, and military logistics and organisation.

But the effect of communication could be felt more broadly. Laurence uses the infrastructure of communication within the Roman province of Britain as a measure of urbanisation and settlement geography. Our understanding of this is important, as Britain lies outside the centre of the empire in the Mediterranean basin, and thus may serve as a guide to the development of communication networks of frontier regions or those newly annexed. Evidence for mobility, Laurence argues, can be used as a measure of cultural change – and this is encouraged by communication with the centre of the empire – the idea of connectivity.

Recent work has illustrated how much the army and its commanders depended on communication networks within and without the empire.[10] Such a reliance clearly extended to logistics.[11] Coulston approaches the issue of military logistics and transport using iconographic evidence. Trajan's Column (Figure 1.2) must be seen as part of the growing imperial necessity to advertise achievement – it was propaganda in its truest form, celebrating the glory of Rome and the emperors – but also of depicting Roman control and influence over the earth.[12] The column represents the first depiction of military transport

Figure 1.2 Trajan's Column. Photograph R. Laurence

in Roman art, but is central to the imperial message that soldiers, and perhaps also the Roman viewer, were the emperor's partner in the conquest of peoples and land. Technical skill is shown as a way of dominating peoples and lands, and this was evident to inhabitants of Rome. The great imperial building projects of emperors like Trajan and Hadrian utilised the resources of the whole empire. In the case of the Pantheon, for example, column shafts of Egyptian granite were transported from the quarries of Aswan and the Eastern Desert (Figure 1.3). In this way the Romans took the monolithic culture of Egypt to a new level, and superseded the pharaohs in their technical abilities.[13]

The chapters here collected discuss themes of central importance to our understanding of the Roman world. These themes are crucial in that they show how mobile a culture the Roman empire was or had a potential to be. In this respect they show how far the discussion of mobility has come from the seemingly entrenched theories of Finley – a static population hemmed into local communities by the high cost of land transport.[14] Rather, horizons had been opened up under the Romans to such an extent that many individuals, not just soldiers or state officials, could make long journeys to all parts of the empire. In the words of Aristeides, every man could go where he wished.

Figure 1.3 Pantheon. Photograph R. Laurence

NOTES

1 Brodersen (1999) and this volume; Gallazi and Kramer (1998).
2 Aristeides of Smyrna, *Or.* XXVI 100–2 (tr. Friedländer). For a discussion of *elogia*, see Friedländer (1913) 268–71.
3 For discussion of a 'civilising ethos', see Woolf (1998) 48–76.
4 See Nicolet (1991) for these ideas.
5 On environmental issues see Hughes (1994).
6 Dilke (1985; 1987).
7 Laurence (1999). On Egypt, see Adams (forthcoming).
8 Pliny the Younger, *Ep.* 2.17.
9 For the link between geography and fiscal administration, see Nicolet (1991).
10 Sherk (1974) initially; Austin and Rankov (1995).
11 Erdkamp (1998); Roth (1999).
12 See generally Nicolet (1991).
13 Adams (2000).
14 Finley (1973).

2

THE PRESENTATION OF GEOGRAPHICAL KNOWLEDGE FOR TRAVEL AND TRANSPORT IN THE ROMAN WORLD

Itineraria non tantum adnotata sed etiam picta

Kai Brodersen

A recent encyclopedia article on land transport in the ancient world[1] surveys the problems associated with travel and transport as realised in classical scholarship. It concentrates on the technicalities of land transport, on harnessing, carts and carriages, on the economic importance of the transport of goods, public transport, travelling, infrastructure and the cost of transport. While this article shows how difficult and expensive land transport was, one problem is not addressed: the simple question of how one knew where to go at all before one even started to travel or transport goods from A to B. It is this – perhaps too obvious – question that we shall try to answer in the following pages: how was geographical knowledge presented for travel and transport in the Roman world?

Successful transport, after all, depends primarily on a knowledge of the world one wants to move in, on the answers to the questions of where one can go and how one can get there. In an age of detailed scale maps (to say nothing about satellite navigation), of geography as a standard school subject, and of modern means of communication, the answers are obvious. But what of Roman times?

THE WANDERINGS OF GALEN

Let us start with the sorry tale of a very learned man in the second century AD, the imperial doctor Galen. He had heard about a special kind of blood-red soil only to be found in Hephaistias on the Greek island of Lemnos, and, as a true scholar, he wanted to do 'hands on' research on this feature personally:

> Of course, I did not hesitate to travel to Lemnos, since I wanted to see how much blood is mixed with the soil. When I was on my second journey from Asia to Rome, travelling by land via Thrace and Macedonia, I first went from Alexandreia Troas towards Lemnos, taking a boat en route to Thessalonike. I agreed with the captain that we would first land at Lemnos.
>
> He did indeed land there, but not where it would have been necessary – for initially I had not known that there are two cities on the island, but had assumed that, like Samos, Chios, Kos, Andros, Tenos and all the other islands in the Aegean, Lemnos as well had only one city by the same name on the island. When I had gone ashore from the boat, I was informed that the city was called 'Myrina' and that it was neither near a (monument for) Philoktetes nor near a hill, which was sacred to Hephaistos, in the environment of this city, but in that of a different one, Hephaistias – and that this city was not close to Myrina.
>
> The captain could not wait for me any longer, so I postponed a visit to Hephaistias until another time, when I would return from Rome to Asia. And this was done by me eventually, as I had hoped and planned. When I had crossed from Italy to Macedonia and when I had travelled through nearly all of it, I got to Philippi, a city in the vicinity of Thrace. From there I reached sea, some 120 *stadia* away, and from there I first sailed to Thasos, some 200 *stadia*, and from there some 700 *stadia* to Lemnos, from there another 700 to Alexandreia Troas.
>
> I have written about my journey and the *stadia* extensively, so that anyone, who might – like me – want to visit Hephaistias, knows its location and can arrange for the journey accordingly. On the whole island of Lemnos there are in the part that verges towards sunrise Hephaistias, and in the one towards sunset, Myrina.
>
> (Galen, *De simplicium medicamentorum temperamentis* 9)

A learned man, working for the imperial court, travels from Asia Minor to Rome and back several times, but finds himself short of even such basic information as to how many cities there are on an Aegean island and how to get to the one he wants to find. He simply does not know it. His school and study subjects, of course, might have included, if any, some 'Homeric' geography of myth and fable, but never any 'real' geography. It was to take another 1,400 years before this was taught at schools at all, by reading 'updated' classical geographers, and 1700 before geography became a standard school subject.[2] And although Galen must have had access to a great deal of learned information, he does not even consider looking the details up on a map (after all, his contemporary, the astronomer Ptolemy, had, as we know, collected quite a number of geographical data which were later used for the production of a map) – this kind of presentation of geographical knowledge simply does not occur to him (and there is no evidence at all that Ptolemy's findings were noticed by his contemporaries, or indeed for another two centuries[3]).

So all Galen can think of is to ask a professional, a captain – who, as it turns out, does not know the details of the geography either and fails to get Galen to Hephaistias. Only after a second, more successful attempt does the doctor write down an itinerary for the benefit of future researchers; he does not assume that they would otherwise have access to this kind of knowledge, and he certainly does not even mention that drawing a map might perhaps be a better alternative of presenting his newly gained geographical knowledge.

BUT THE 'PRACTICAL' ROMANS 'MUST HAVE HAD' MAPS TO SCALE

Now, of course, this lack of maps cannot be true, or so we are told by modern historians. The Romans, after all, were such a practical people, weren't they? 'Considering their highly developed administrative abilities, it is hard to believe that maps were not in common use.'[4] Indeed, any modern historical atlas will show us scale maps of 'the world the Romans knew'. And since 'it may seem surprising that scale maps were virtually unkown in medieval Europe in view of the extremely competent and complex scale maps produced by the Roman surveyors from the first to the third century AD', the only explanation brought forward for this typically dark medieval lack of good maps is 'that the Roman tradition of scale maps died out altogether.'[5] So it is considered inconceivable that 'the Romans' in general, and learned high-ranking officials like Galen in particular, travelled without access to detailed geographical knowledge available on maps. Already 'by the first century BC the knowledge of maps had been very widespread in Rome', asserts one scholar,[6] and another takes the title of the *comes formarum* as evidence 'for what must have been a civil service maps and plans department'.[7] In fact, the *comes formarum* was only in charge of the aqueducts, as was demonstrated by Mommsen more than a century ago[8] – and there is precious little evidence for a 'Roman tradition of scale maps', or, to be precise (as I have argued in detail elsewhere[9]), no evidence at all: not even the surveyors' maps are drawn to scale.

So to substantiate this claim for the question of this chapter let us look at the evidence produced for 'missing links' between ancient geographical knowledge and the assumption that maps to scale have always (and certainly in classical times) been the best, even the only reasonable way to represent this knowledge.

1. 'The earliest Greek map to come down to us in any form and the first physical relief map known' was 'discovered' by A.E. Johnston on coins from fourth-century BC Ionia.[10] The reverse of some of a series of tetradrachms show an irregular pattern, changing from die to die, with what looked to Johnston like 'a map depicting the physical relief of the hinterland of Ephesus, an area of approximately 90 square miles'. The features 'must have been an attempt to show surface phenomena, almost certainly vegetation' (note the 'must have been' again) and are thus 'remarkably close to that of a modern map'. Similarly, O.A.W. Dilke has stated that these coins 'clearly portray, though only in rough form, some sort of relief map of the hinterland of Ephesus'.[11] However, the pattern changes significantly from die to die (which would be rather problematic if this really were a map to scale), and the details are indeed far too 'rough' to allow any interpretation of this kind. It is not surprising, then, that even the author of this interpretation has since become 'less sure of her theory',[12] which should not now be repeated in books on the history of geography and cartography.[13]

2. Both the standard textbook on 'Greek and Roman Maps'[14] and the monumental first volume of the 'History of Cartography'[15] present us with a Roman 'Map of Gaul', a block of sandstone which can be tilted upwards and then looks a bit like an outline, to scale and with north at the top, of modern France. In addition, three of the many holes are taken to represent 'the Gallic religious centers of Puy de Dôme, Autun, and Grand' – which leads to the rash conclusion that the stone dates from the time of Caesar's Gallic war.[16] The 'map' was a stray find, made – or so it is reported – in 1976 by one Pierre Camus in a Roman camp in Mauchamp on the Aisne. It has never been properly published; the details of where the stone was found (and where it is now) are obscure, and the proposed first-century BC date is completely unsubstantiated. So great, however, is the desire to find the 'missing link' that the discovery of this 'Map of Gaul' was specifically greeted as a 'besonderes Finderglueck' (special discoverer's luck)[17]: here, so it was claimed, is 'one of the few surviving geographical maps of the pre-Augustan period'.[18]

However, it has recently been shown[19] that the reported date of the find, 1976, is at odds with the fact that the discoverer, Pierre Camus, has used just this 'map' on a drawing of the cover of a historical novel which he published in 1974.[20] Given this oddity, and the fact that there is no independent evidence for the find in 1976 at all, let alone for the antiquity of the 'map', all this stone proves are dangers for scholarship. This was an example of the fervent quest for the Roman map allaying any reasonable caution: a stone presented as 'ancient' was accepted and became a pivotal piece of evidence for 'the Roman tradition of scale maps'.

3. In the summer of 1999, the French magazine *Archéologie* reported the discovery of an antique mosaic in a rich third- to fourth-century villa in Ammaedara, modern Haidra, in the North African hinterland, some 160 miles

Figure 2.1 Ionian coins with 'geographical representations'
Source: Head 1892

south-west of Carthage.[21] In a graphic display which resembles a map, more than a dozen Mediterranean islands are presented, including Cnidus, Cyprus, Egusa, Naxos, Rhodes, Scyrus and – too late for Galen – Lemnos. All these islands vignettes show one important harbour city each (compare Galen's claim that all Aegean islands 'have only one city by the same name on the island') and identify the location by writing down the name of the place. This is important, as the outlines of the islands and their location in the sea are not by themselves recognisable, not drawn to any scale and not representing the 'real' geographical position in the Mediterranean. Indeed, not even all of them are islands (Cnidus, for example, represented here as an island, is a peninsula). And there are three separate 'islands' named Cyprus, Idalium and Paphus – where we know, of course, that the latter two are cities on the first. Did the artist, like Galen, believe in the 'one island–one city' theory so much that this was the only solution he could think of?

So again, while the graphical representation reminds us of a 'map', the mosaic does not provide evidence for a 'Roman tradition of scale maps'. The archaeologist who published the find rightly assumes that it is rather a graphical representation of an itinerary, of stations along the route of a mythical journey of Venus, the *navigium Veneris*. Geographical realities did not matter here, and elsewhere. Neither this map nor any source which the artist might have had access to was useful for travel and transport at all.

4. Perhaps a century later than the Ammaedara mosaic are the graphical representations of islands – like Britain – and regions in the *Notitia Dignitatum*. Again, these graphics remind us of maps, but – like on the mosaic – the outlines of the islands or regions are completely arbitrary, as is the position of the 'cities', again marked by vignettes, on them: 'The presentation is purely conventional and the illustrations were not intended to be maps . . . The disposition of the names on the "island" is conventional rather than geographical.'[22]

5. Finally, scholarly literature on geographical knowledge often refers 'to the best known example of Byzantine cartography,'[23] the sixth-century 'Madaba Map', a mosaic in a church in Madaba (Medaba) in what is now Jordan. This map, discovered in 1896 and preliminarily published by Avi-Yonah in 1956, still awaits final publication.[24] The mosaic represents the Holy Land from the

Mediterranean shore to the desert and from Tyrus and Sidon to the Nile delta; Jerusalem is – of course – at the centre of the image, while cities and stations – 157 of them are preserved – are marked by vignettes and individually named; most of them are obviously taken from Eusebius' catalogue of biblical toponyms and from an itinerary, a list of road stations, in the Near East; roads, however, are not marked on the mosaic. The graphic is not at all to scale: the coastline from the Nile to northern Syria is a more or less straight line, ignoring the fact that the first portion runs from west to east, the second from south to north, and the 'scale', if one were to measure it,[25] oscillates between 1:15,000 and 1:600. So the 'Madaba Map' cannot be used as evidence for a 'tradition of scale maps', let alone for a useful representation of geographical knowledge. Rather, if the Ammaedara mosaic served a 'pagan' interest in myth, the Madaba mosaic fulfils a similar purpose in a Christian context – its subject is not useful geographical knowledge for people 'on the road', but a graphical representation of places important for the reader of the Bible.

To sum up, all five artefacts, which have been adduced by scholars so far as material evidence for a 'Roman tradition of scale maps', fail to prove the existence of this tradition, and resorting to the 'must have been' variety of logic does not solve the problem of how geographical knowledge was presented.

ITINERARIA ADNOTATA (ANNOTATED ITINERARIES)

However, there were other means of representation: itineraries, list of stations along a route, of the kind which Galen found most useful for future researchers, should they want to travel to Hephaistias on Lemnos. There is ample evidence for the use of itineraries,[26] not least in military action. Here is how the Roman emperor Alexander Severus enabled his troops to find their way:

> The dates of the *itinera* were publicly displayed; two months before the event he published an *edictum*, in which was written: 'on that day, at that hour, I shall go forth from the city and, if the gods allow it, I will stay in the first station', detailing then the stations one after another, then the camps, and then where provisions are to be had, and all that for as long as one arrived at the barbarians' borders. From there everything was silenced, and all went without certainty, lest the barbarians would know the Roman plans.
>
> (Historia Augusta, *Alexander Severus* 45, 2–3)

How common this kind of organisation of geographical knowledge was is evident also from a reference in an unusual context for military questions, St Ambrose's commentary on Psalm 118 (119 in the King James' Bible), verse 33 ('Teach me, O Lord, the way of thy statutes, and I shall keep it unto the end'):

> A soldier who starts an *iter* does not by himself arrange for the order of the march, does not take a way by his own judgement and does not take an arbitrary shortcut, lest he leaves the *signa*, but he receives from the *imperator* an *itinerarium*.[27]
>
> (St Ambrose, *Expositio Psalmi* CXVIII 5, 2)

We do, in fact, have a number of Roman itineraries, first and foremost the well-known literary *Itinerarium Antonini*, which gives us an idea of what these itineraries looked like: simple lists of stations along a route, giving their names and the distances between them. Similar lists are also known from inscriptions, for example, a large octangular block from Tongeren (Tongres) in Belgia, dating to *c.* AD 200 and detailing lists of stations and distances (in the Gallic unit of *leugae*) along several routes meeting at the place where the stone was posted. Here is what the three legible sides preserve (giving modern placenames):

Cologne 11 leugae
Bonn 9 leugae
Remagen 8 leugae
Andernach 8 leugae
Koblenz 8 leugae
Boppard 8 leugae
Oberwesel 8 leugae
Bingen 12 leugae
Mainz 9 leugae
Nierstein 11 leugae
Worms 11 leugae

———

. . . 15 leugae
Warcq 15 leugae
La Noue Maga [?] 12 leugae
Reims 12 leugae
Fismes 12 leugae
Soissons 16 leugae
Pontoise 9 leugae
Roye 8 leugae
Peti-Hangest .. leugae
Amiens .. leugae

———

equally from Cassel
the region of the Atrebates 14 leugae
Arras .. leugae
equally from Bavay

(*CIL* XIII 9158 = XVIII 2, 675)

Similar epigraphical lists[28] are preserved from the same region in Allichamps (*CIL* XIII 8922 = XVII 2, 489), Autun (*CIL* XIII 2681 = XVII 2, 409) and Junglinster (*CIL* XIII 4085 = XVII 2, 676), but also from the African Fedj-Souioud (*CIL* VIII 10118 = 22247).[29] The latest discovery of such a list of places and distances was made in Lycia: the so-called 'Stadiasmus' from Patara (*SEG* XLIV 1205), as yet only known from a preliminary publication and some recent photographs (see Figure 3.5 below).[30] Apparently, here the formation of the Roman province of Lycia was marked not by the public display of a map or anything maplike, but a monumental inscription with the names, and distances, of important places in the new province.

ITINERARIA NON TANTUM ADNOTATA SED ETIAM PICTA (ITINERARIES NOT ONLY ANNOTATED BUT ALSO ILLUSTRATED)

Presenting geographical knowledge in this form of itinerary was obviously a very successful and practicable method, and evidently useful for travel and transport. It did indeed successfully solve the problem not addressed in the article referred to at the beginning of the essay: how do I get the goods – or myself – from A to B without access to scale maps.

However, this is not the whole story. The Roman military writer Flavius Vegetius Renatus *c.* AD 400 recommends in his work *De re militari* the use not only of annotated, but also of 'painted' itineraries: '*itineraria provinciarum, in quibus necessitas gerebatur, non tantum adnotata sed etiam picta* [itineraries of the provinces in which the emergency occurred not annotated but illustrated]' (Vegetius 3, 6).

Given the search for the 'missing link', it is not surprising that the Latin term *itineraria picta*, known only from this passage, has been translated as 'coloured maps'[31] or 'proper maps'.[32] But as we have seen, there is no other evidence for 'proper maps', and there is no need to translate *itineraria picta* as such. What then could an *itinerarium pictum* have looked like?

There are a number of artefacts where a textual itinerary is adorned by graphical elements: the four second-century AD Vicarello goblets, for example, list the stations, and distances, between Cadiz and Rome, on silver beakers in the shape of a Roman milestone (*CIL* XI 3281–4). Two vessels, one from Rudge Coppice near Froxfield in Wiltshire (*RIB* II 2, 2415.53) and one from Amiens (*AE* 1950, 56), set a list of stations on Hadrian's wall very similar to the one in the Ravenna Cosmography next to images of turrets or crenellations:

Mais Aballava Uxelodunum Camoblannis Banna Esica
Bowness-on-Solway, Burg-by-Sands, Stanwix, Castlesteads, Birdoswald, Great Chesters

Visually more attractive is the list of stations on the so-called Dura Shield, a third-century AD leather fragment from Dura-Europas found in 1922 and published soon afterwards (Figure 2.2).[33] The centre of this round leather patch was covered with the images of four boats and surrounded by a circular line, along which we find vignettes of the sort familiar from the mosaics, though less elaborate, the names of stations and the distances between them:

River Panysos .. miles
Odessos (Varna) .. miles
Bizone .. miles
Kallantis (Mangalia?) .. miles
Tomis (Constanza) 33 miles
(Istropolis on the) Istros 40 miles
Danubios (Danube) .. miles
Tyras (Belgorod) 84 miles
(Olbia on the) Borysthenes (Nikolaev) .. miles
Chersonesos Taurike (Crimea) .. miles
Trapezus mountain (Krimskie Gory) .. miles
Ardaba (or Arta) .. miles

(Dura parchment 9)

The preserved portion of the shield, then, presents stations along one coastal route (road) of the northern part of the Black Sea (there are no crossroads); as

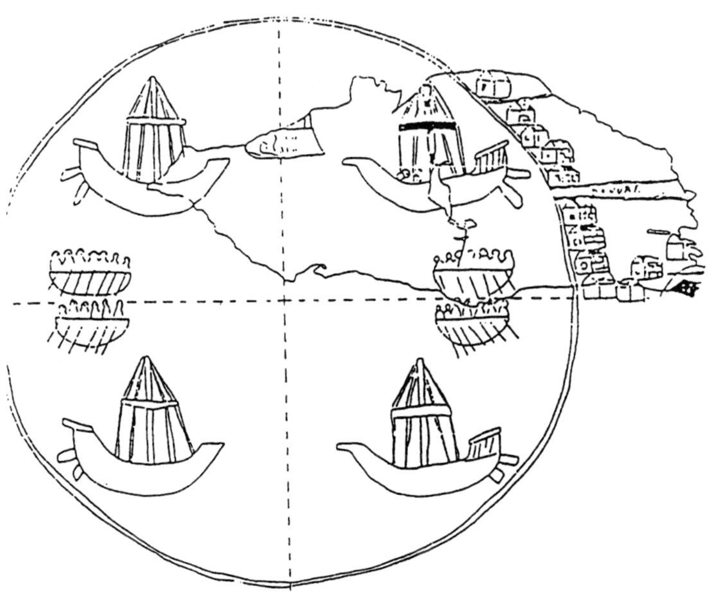

Figure 2.2 Dura-Europas Shield © Kai Brodersen

less than a sixth of the circle is preserved, it can be argued that the original was not limited to stations on the Black Sea coast, but possibly covered the eastern (or even whole) Mediterranean. In any case it is far from an accurate representation of geographical realities (let alone scale): what is in reality a rugged and complicated coastline is represented here as a full circle – a decorative form of *itinerarium pictum*, but not a map.

THE LONDON TUBE DIAGRAM AND ITS ANCIENT PREDECESSORS

There is now even more to ancient *itineraria picta* than the Dura Shield: on 6 December 1999 an article by David Keys in the *Independent* reported: 'Scholars uncover oldest map in Western World' (there is a garbled and unacknowledged fake of this article in the German paper *Frankfurter Allgemeine Zeitung* of 8 December 1999, which completely misunderstands Keys, but at least proves how badly news travels even in the age of electronic communication). Not quite: what has been discovered, and published – so far only in part – is what I take to be an *itinerarium pictum*.[34]

The papyrus, apparently written in first-century BC Antaiupolis in Upper Egypt (the provenance is not proven, and the papyrus is now in a private collection), contains, *inter alia*, a number of highly artful portraits and details of human bodies, studies of animals – birds, fish, and mammals, including a tiger and a giraffe – as well as fantasy beings like a star-dog and a panther-crocodile. It also preserves a more complete text of a geographical work by the early first-century BC geographer Artemidorus that was previously known from quotations by later authors;[35] the portion published so far deals with Spain[36] – and is accompanied by a drawing of what the editors interpret as a representation of Spain.

This drawing contains a number of intersecting lines obviously representing roads or rivers, and what look like vignettes of the sort familiar to us from the (later) Dura Shield as representations of stations. It does not, however, contain any text, and the editors have argued that the project was abandoned before the names of the stations, and possibly the distances, were added. The whole drawing is rather unalluring, especially when compared to the fine images of the animals – an interesting comment on the state of the art of mapping in the time and country of the papyrus: a draughtsman, who could draw beautiful portraits, elaborate images of fish and fanciful representations of star-dogs, could not think of anything but a schematic diagram of routes and stations when confronted with the task of illustrating a geographical text. What he produces, then, is a diagram, probably correct in the representation of 'topological' relations of spaces (of the sort familiar from the London Tube diagram), but certainly not to scale.

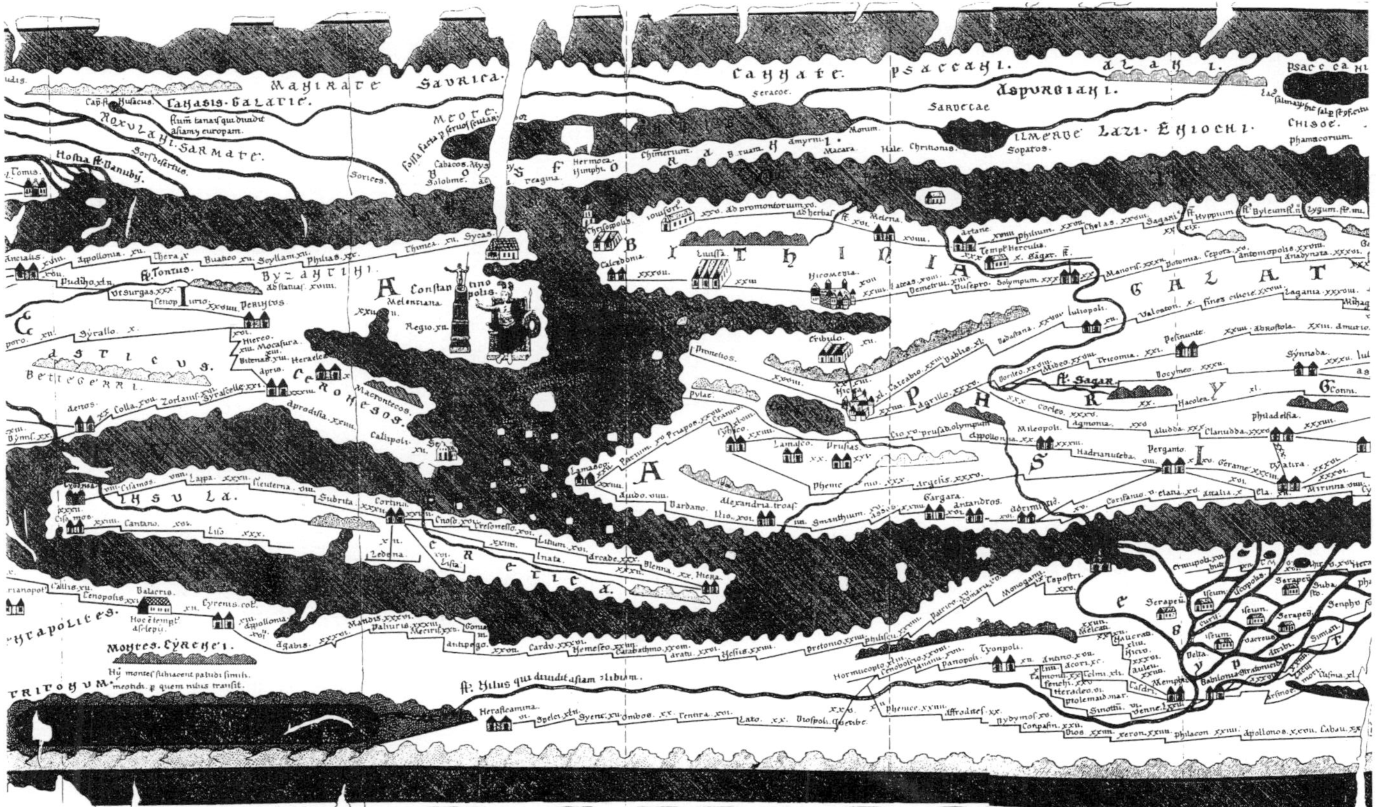

Figure 2.3 Peutinger Table: the eastern Mediterranean, including Crete, the Nile delta, Constantinople and modern Turkey (With permission from the Österreichische NationalBibliothek.)

In fact, a similar diagram dating to at least Roman imperial times has been known for a long time, if only from a twelfth-century copy: the so-called *Tabula Peutingeriana*, a parchment roll of *c*. 30 cm height and nearly 700 cm width, representing routes through the ancient world from Britain (this portion was lost, though) to India. The age of the Roman original of the *tabula* is unclear: it does list monuments of Alexander the Great (fourth century BC) and Pompeii (destroyed in AD 79), but it also shows Constantinople (founded in the 4th century AD) and a number of pilgrim's stations like the Mount of Olives in Jerusalem and St Peter's in Rome. The new papyrus suggests that the idea of representing space in a diagram was familiar already in the first century BC.[37] As on this papyrus and on the Dura Shield, stations along the routes on the *tabula* are marked by hooks or vignettes of stations, and, as on the Dura Shield, the names of the stations and the distances are spelled out and cannot otherwise be deduced from the image; there is certainly no concept of scale. What is more, geographical accuracy is not sought, as the editor of the magnificent facsimile has stated: 'The courses of rivers, if they correspond to real watercourses at all, are far removed from their proper form; they are crossed by roads which in reality run elsewhere altogether, and thus suggest that they were often only drawn so as to enliven the image optically; the same is generally true for the mountain ranges.'[38]

What is generally correct, however, is the series of stations along a route, and the distances written along the lines representing the roads in between these stations, and it is safe to assume the same sort of only 'topological' accuracy in the relative position of stations to each other for the unidentified stations on the Artemidorus diagram.

To be sure, such diagrams are a very successful way of organising and presenting geographical knowledge – as was discovered in London in the 1930s: Henry (Harry) Beck, a part-time draughtsman for the London Underground Group, suggested in 1931 replacement of the geographically accurate scale maps used as a representation of the network with a radically simplified topological diagram, which gives up the concept of scale altogether and presents all stations as more or less equidistant along routes which know only angles of 90 or 45 degrees. Initially, Beck's suggestion was rejected by his superiors, but two years later a trial run of some leaflets with Beck's diagram proved to be such a success with the travelling public that it replaced all maps[39] – and has since become the standard representation of regional transport all over the world (see Figure 2.4), for the simple reason that it organises geographical knowledge in a visually attractive and practical way.

THE PRACTICAL ROMANS, AFTER ALL

To sum up: the question of how, in Roman times, geographical knowledge was presented to those engaging in travel and transport cannot be ignored. There

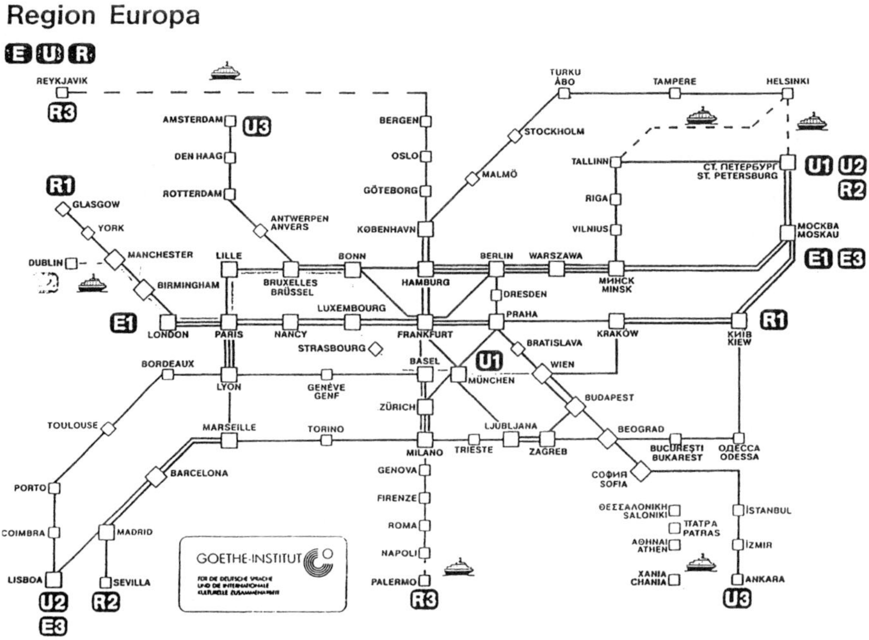

Figure 2.4 The current map of European communications. (Courtesy of B. Luley, Goethe Institute, Munich.)

is no obvious answer, either, as the claim that there simply 'must have been' maps to scale for that purpose seems to be wrong.

This, however, does not make the Romans look unpractical: it is certainly no coincidence that the highly redundant information on scale maps cannot even now be read by everyone and that modern 'routing' software, which translates the information stored in maps into simple lists of stations and distances, is presently such a commercial success.

Rather, geographical knowledge was organised, and presented, in itineraries. The risks of this method are small (you might admittedly fail to realise that there is more than one city on an island), the gains, however, great: an itinerary allows you to plan travel or transport from A to B successfully, and it is this method which was adopted throughout antiquity. Simple *itineraria adnotata* are enough if there is only one route, but if there is a choice of routes to be taken, the ideal is – from the first-century BC Artemidorus papyrus to the London Tube diagram – an *itinerarium pictum*.

NOTES

1 Raepsaet (1999).
2 Cf. Brodersen (1996).
3 Polaschek (1965) 764–5.
4 Moreland and Bannister (1989) 4.
5 Harvey (1987) 466.
6 Sherk (1974) 559.
7 Dilke (1987) 244; note the 'must have been'.
8 Mommsen (1887) 1047; cf. Demandt (1989) 383.
9 Brodersen (1995). Salway in this volume attacks my scepticism on the so-called 'map' of Agrippa in the Porticus Vipsania at Rome, which I take to be a monumental inscription, as 'unwarranted on a number of counts', going on to mention two (and failing to mention that I discuss all of them): first, the 'venerable tradition associating pictorial illustration with public porticoes' as witnessed by the will of Theophrastus – on which see Brodersen (1995) 75, arguing that this will rather shows how rare and valuable such depictions were; second, the wording of Pliny, *NH* 3, 17 'spectandum' – a word Pliny uses to refer to texts elsewhere, cf. Brodersen (1995) 277–278 – and two late antique passages – Eumenius, in AD 297, referring to an idea he has for the school in Autun, and *Anth. Lat.* 724, referring to a redrafting of the anonymous *Divisio orbis terrarum* in AD 435 – on both see Brodersen (1995) 106–7. I fail to see how texts that do not refer to the Porticus Vipsania or Agrippa at all, and that are more than three centuries apart from it, are able to prove Salway's point, while direct references to the Porticus Vipsania never mention the 'map': cf. Martial, Epigr. 1, 108, 3; Tacitus, *Hist.* 1, 31, 2; Plutarch, *Galba* 25, 9; Dio Cassius 55, 8, 3–4, etc. – all discussed in Brodersen (1995) 275, and not discussed by Salway in his article. To end on a conciliatory note: we both agree, of course, that whatever the portico displayed cannot be regarded, in Salway's words, as a 'plausible archetype for the lists in the Antonine Itineraries or the routes on the Peutinger Table and general public knowledge thereof'.
10 Johnston (1967).
11 Dilke (1985) 146.
12 Dilke (1988) 92.
13 But see, e.g., Olshausen (1991) 91.
14 Dilke (1985) 102–3.
15 Harley and Woodward (1987) 206–7.
16 Dilke (1987) 207.
17 Hinrichs (1987) 666.
18 Dilke (1987) 207.
19 Brodersen (1995) 143–4.
20 Camus (1974); I have used the copy in the BN Paris (8° 50383).
21 Bejaoui (1999); cf. now Bejaoui (1999/2000) and Brodersen (2001).
22 Jones and Mattingly (1990) 33.
23 Dilke (1987) 264.
24 Avi-Yonah (1956); a final publication was promised by Donner and Cueppers (1977), but has not appeared; cf. meanwhile Piccirillo (1989).
25 Avi-Yonah (1954) 21.

26 On itineraries in general cf. Brodersen (1995) 165–90 and Salway in this volume.

27 Salway in this volume argues that this *itinerarium* is not a real thing, but a 'construction of the reader' – a view that seems to me unconvincing in view of the actual text of St Ambrose (*itinerarium ab imperatore accipit; praescripto incedit ordine; a praescripto itinere*).

28 Salway in this volume introduces the term *tabellarium* for these lists – a term for which, in this meaning, there is no ancient evidence at all (whatever *CIL* I^2 638 refers to with *tabelarios*, it cannot be the accusative plural of *tabellarium*, and none of the other lists, or references to lists, uses the term) and for this reason should not be used.

29 There is no evidence for the assumption that the *miliarium aureum* in Rome carried such an inscription; cf. Brodersen (1996/7).

30 Sahin (1994); photos in Isik (1999).

31 Milner (1993) 17.

32 Stueckelberger (1994) 69: 'eigentliche Karten'.

33 The principal publication is Cumont (1926).

34 Gallazzi and Kramer (1998/9); cf. Brodersen (1999).

35 Brodersen (1997), published before the discovery, obviously reflects only this earlier knowledge.

36 On this see now Kramer and Kramer (2000), who extensively mock the article on *Hispania* in *Der Neue Pauly* as not well-enough researched – a deficiency true for their own article as well: they name a wrong author for the article they poke fun at (K. Brodersen instead of P. Barcelo).

37 See Brodersen (forthcoming (b)).

38 Weber (1976) 12.

39 Garland (1994).

3

TRAVEL, *ITINERARIA* AND *TABELLARIA*

Benet Salway

The number of surviving itineraries, listing the names of places and the distances between them along a given route, leaves no doubt that they were a feature of the Roman world, and that they were of use to ancient travellers in the course of actual journeys. Indeed, the manuscript itineraries have for over a century played an important role in a debate over the extent of geographical knowledge available to the Roman public and the nature of their understanding of physical space. However, no clear scholarly consensus as to the significance of these documents has emerged, while their relationship with the epigraphic material remains relatively unexplored.[1] It is my intention, through a careful reconsideration of the manuscript itineraries and of the testimony of one notorious epigraphic document, the so-called Elogium from Polla, to give identity to a particular class of epigraphic text and explore its relationship with the manuscript itineraries, in order to demonstrate the likely regularity of its role in mediating itinerary information to Roman travellers.

ITINERARIA

Among the written itinerary documents that have come down to us via the medieval manuscript tradition, the most extensive and well known are, first, the third-century collection of provincial and maritime itineraries that circulated under the name of an emperor Antoninus (the *Itinerarium Antonini*, Figure 3.1), known from over twenty copies, and, second, the so-called Peutinger Table (*Tabula Peutingeriana*, Figures 1.1, 2.3 and 8.3), a unique document of somewhat later date.[2] This latter is a parchment roll, comprising eleven (originally twelve) leaves, in total 6.75 m. long, by a modest 32–4 cm. high, dimensions that are in keeping with an antique papyrus archetype. It bears a multicoloured depiction of the entire inhabited world (*oecumene*) as the Romans conceived it, albeit necessarily foreshortened (north–south) and elongated (east–west) and carefully planned so as to put a seated personification of Rome

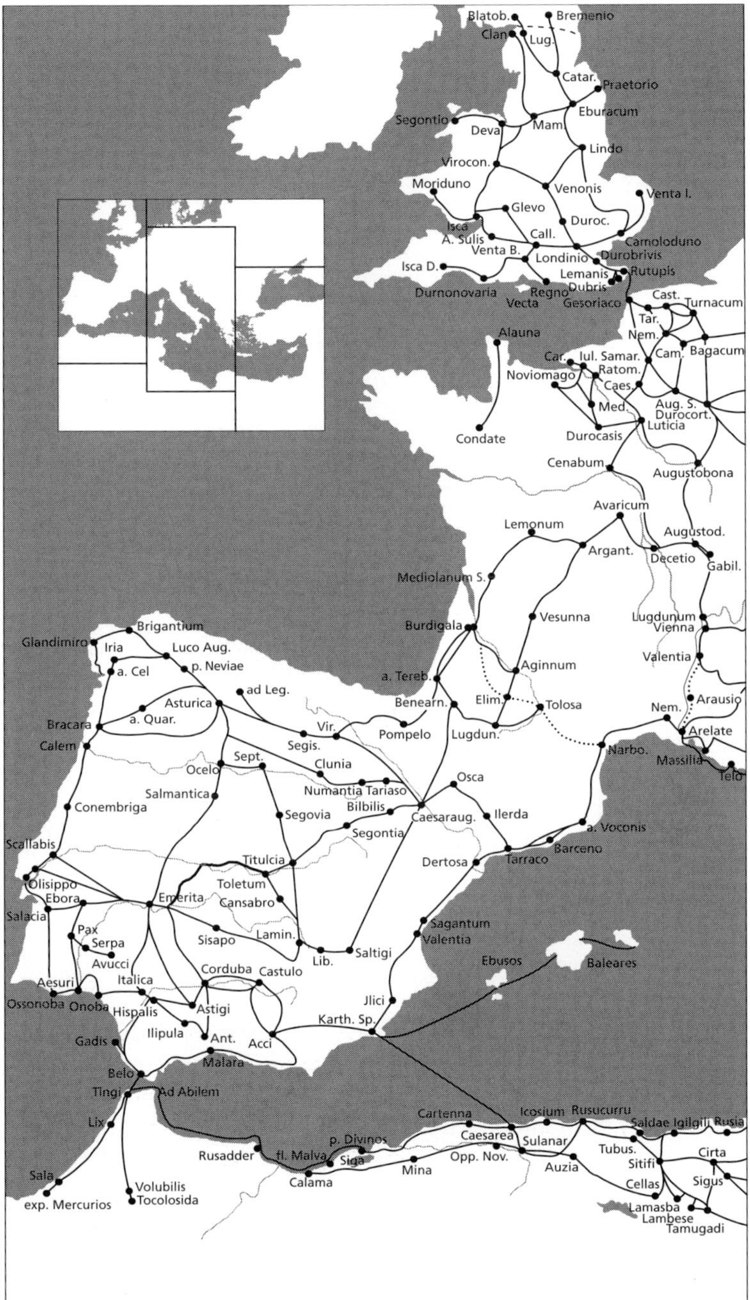

Figure 3.1 Coverage of the empire by the Antonine and Bordeaux Itineraries (after Cuntz)

continued

Figure 3.1 continued

Trosmis
Salsovia
Axiup.
Dorostoro
Tomos
Oesco
Novas
Marcian.
Odisso
Trapezus
Zigana
Cabile
Anchialis
Beroa
Develto
Philippopoli
Bessapara
Hadrian.
Ostudizo
Bergule
Olot.
Nicopoli
Satala
Sebastia
Sebastop.
Zara
Zimara
Claud.
Gratia
Chalcedonia
Nicomedia
Tavia
Philippis
Traian.
Apris
Byzantium
Heractea
Nicaea
Dorilao
Ancyra
Tonosa
Callip.
Lampsacum
Germa
Ptandari
Melitena
Cod.
Caesarea
Ilio
Pessinunto
Parnasso
Comana
Arabisso
Cocuso
Perre
Samos.
Lemnos
Adramyttio
German.
Doliche
Edissa
Pergamo
Thyatira
Archelais
Tyana
Lesbos
Anaz
Car.
Euboea
Sardio
Nic.
Zeugma
Chalcis
Hierapoli
Mopsucrenae
Cyrr.
Laudicia
Tarso
Aeg.
Hierapolis
Athenis
Icaria
Alex.
Antiochia
Calcis
Barbal.
Delos
Myconos
Androna
Seriane
Laodicia
Ap.
Ep.
Antaradus
Hemesa
Cyprum
Eumari
Tripoli
Hel.
Beryto
Damasco
Tyro
Neve
Caesarea
Scytopoli
Neapolim
Diospoli
Hiericho
Ascalo
Hierusolyma
Hebron
Darnis
Pelusio
Rinocorura
Alexandria
Cyno
Taposiris
Catabethmon
Paraetonio
Andro
Thou
Serapiu
Heliu
Clysmo
Babylon
Memphi
Afrodito
Hipponon
Oxyrincho
Hermupoli
Antinou
Lyco
Pano
Ptolomais
Copton
Tentyra
Thebas
Lato
Apollono. S.
Ombos
Beronicen
Syene
Talmis
Hiera Sycaminos
Contra Pselcis
Scale 1:10 000 000
0 50 100 200 300 400 500
Milia passuum (1 = 1480 m)
0 50 100 200 300
Leugae (1 = 2220 m)

in a nimbus at the centre both horizontally and vertically. The outline of the land masses with their major mountains and rivers, and with the Mediterranean sandwiched between, is clearly recognisable. Nevertheless, the dominant feature is the representation of the network of land routes by a diagram of angular lines, on which the distances between named places are indicated and significant locations are picked out by little icons or vignettes. We also possess an increasing number of inscribed itinerary lists in various durable media, ranging from the portable to the monumental.[3] Besides these, there survive several itineraries relating to specific travels: a broken plaque from the columbarium at the Vigna Codini, between the via Appia and Latina to the south-east of Rome, records eight days' worth of a journey from Cilicia through the Taurus mountains to Cappadocia; from among the papers of one Theophanes from Hermopolis Magna in Egypt survive the accounts of both his outward and return itineraries to and from Antioch *c.* AD 320; and, from the manuscript tradition, the anonymous Bordeaux Itinerary (*Itinerarium Burdigalense*) recounts a pilgrimage from Gaul to Jerusalem undertaken in AD 333.[4] Although written in Greek, Theophanes' itinerary is a document of Roman travel, not only chronologically but culturally, since the distance figures are given in *milia* (sc. *passuum*), that is Roman miles.

The practical utility to a traveller of a written itinerary, or collection of itineraries, would seem to be self-evident and is discussed elsewhere in this volume by Kai Brodersen. Nevertheless, while Greek antiquity has bequeathed us a rich tradition of maritime itineraries (*períploi*), if there existed any organised collection of land itineraries in the mainstream of Hellenic culture, none has come down.[5] The land itinerary seems to be a primarily Roman phenomenon; though in the first century AD Isidore of Charax, a Hellene working within the other great land empire, did produce the *Stathmoì Parthikoí*, listing stations across Mesopotamia and Persia, and measured in *schoînoi* rather than Greek *stádia*.[6] In the absence of Greek precedents, origins for both the Antonine Itinerary and Peutinger Table have been sought variously (i) in the supposed 'World Map' of Agrippa of the Augustan period, in edicts ordering preparations for the imminent journeyings of third-century emperors or, closely connected, (ii) in the files of the administration of the *cursus publicus* (state courier service) and/or the *annona* (military provisioning), or conversely (iii) in a private collection intended for commercial travellers.

The surviving Roman itineraries certainly do exhibit a fairly uniform aspect. They tend to be headed by an indication of the departure and terminal points of the particular *iter* (journey, route), often including a figure for the total distance. The intervening stages are then listed with figures for the distances between them – for most of the empire in Roman miles, with the partial exception of Gaul where leagues (*leugae*) were also in official use. Thereafter the beginning of a new route is frequently introduced by the conjunction *item* (similarly): for example, '*A Terracina Benevento . . . Item a Terracina Neapolim*' (*Itin.Ant.* 121, 8; 122, 4). Unlike Ptolemy's geographical tables, in which the places are listed in the nominative (as is appropriate to the labelling of a map), the itineraries present the intervening stages in a sometimes disconcerting variety of oblique cases (accusative, locative, or ablative), expressing direction 'to' or 'from', in accordance with the syntax of the list as a sentence including an understood verb of motion.[7] For example, from the African section of the Antonine Itinerary (77, 4–78, 3):

Item a Thelepte Tacapas m.p. CLII:

Gemellas	*m.p. XXII*
Gremellas	*m.p. XXV*
Capsae	*m.p. XXIIII*
Thasarte	*m.p. XXXV*
Aquas Tacapitanas	*m.p. XXVIII*
Tacapas	*m.p. XVIII*

This is more clearly demonstrated by reference to the late antique itinerary from Gades (modern Cadiz) to Constantinople preserved in the prelude to the *Chronicon Albeldense* of 883. Here the underlying grammar is made explicit since, in characteristic romance fashion, it is helpfully spelled out with prepositions:

De Gadis {sic} usque ad Cordobam mill. CC
de Cordoba usque ad Toletum mill. CCXX
de Toleto usque ad Caesaraugustam mill. CCC
etc.[8]

As illustrated here itineraries exist at varying levels of detail but what is common, as is also clear from these short extracts, is their laconic nature. There are no geographical details or indications of compass direction. Each itinerary restricts itself to describing the stages along a given route, so that generally junctions with other routes or side roads do not warrant mention. Indeed, it is to be remembered that the itineraries primarily represent routes of travel, not physical roads; they cannot automatically be assumed to map perfectly the Roman public road network (see Figure 3.1).[9] It is, none the less, true that the Italian section of the Antonine Itinerary accords the republican roads radiating from Rome their time-honoured names. Still, even here, where the road names would facilitate cross-referencing, it is only the entries for the Latina and Labicana that are informative in this manner, since they end respectively 'Compitum (i.e. Junction) 15 m.p., enters into the Labicana' and 'to Beneventum by the stages that are on the Praenestina' – though, in fact, modern scholars would consider the road designations confused.[10] The restriction of this phenomenon to this section is explained by the absence of a comprehensive system of universally recognisable and unique labelling for the physical roads of the empire. Accordingly a Roman's answer to the question 'How do I get from London to Oxford?' would not be 'Take the M40' but 'Travel by way of Uxbridge, High Wycombe, and Thame.' The passage into contemporary common parlance of such code numbers is, after all, in part a consequence of the existence of a vast travelling public who regularly need to find their way rapidly through unfamiliar territory.

To the compiler of a Roman itinerary a crossroads or junction that did not coincide with a useful staging point did not merit comment in a system of route description whose objective was the outlining of potential journeys between two fixed points. Nevertheless, an alert reader of the Antonine Itinerary might be tipped off by the toponym 'Bibium' (i.e. *biuium*: the fork), on the road from Aquileia to Siscia (274, 3) to the existence of a junction. Still, it conveys nothing as to the alternative destination (probably Emona, modern Ljubljana). In fact, the only explicit reference to a turn-off – to Bellunum (*uiam Belloio* [sic]: 276, 2) on the route from Aquileia to Lauriacum on the Danube – is so uncommon that it initially caught out the scribe of the seventh-century Escorial manuscript.[11] A further corollary of organising itineraries as a series of routes from A to B is a high degree of needless repetition of material in the coverage of a region. Such clusters have been noted in the Antonine Itinerary around Milan, in Britain, and in Africa, and similar clusters exist around Cocusus in Cappadocia and Edessa in Mesopotamia.[12] Generally a junction or coincidence of route can only be deduced by observation of the appearance of the same location in more than one itinerary. This drawback would only be overcome by the depiction of the routes as a network diagram, as was achieved in the Peutinger Table.

THE NATURE OF THE *TABULA PEUTINGERIANA*

Konrad Miller maintained that the archetype of the Peutinger Table was a pictorial *cosmographia* compiled by one Castorius because this authority is repeatedly cited by the anonymous author of the eighth-century *Ravenna Cosmography*, whose content is closely related to that of the Table.[13] In reality 'Castorius' is just one of the creations of the pseudo-scholarship of the Anonymus Ravennas, whose supposed geographical *philosophi* improbably include Libanius and the consuls of AD 355 (Lollianus and Arbitio). These authorities were invented to hide the fact that the extent of his research was to read off a sister manuscript of the *Tabula Peutingeriana* itself.[14] Castorius may be bogus, but another map of the *oecumene* has been claimed as the common ancestor of both the *Itinerarium Antonini* and *Tabula Peutingeriana*, namely the so-called 'Map of Agrippa' on the Porticus Vipsania at Rome. Although not the first to suggest this idea, Wilhelm Kubitschek vigorously promoted it in several studies, including a number of articles in the Pauly-Wissowa *Real-Encyclopädie der classischen Altertumswissenschaft*. Assuming the map to have depicted the road network, the information was claimed to be derived from the archives of the administration of the *cursus publicus*, which Suetonius asserts was established by Augustus (*Aug*. 49.5). The Antonine Itinerary was thus understood as having been excerpted from this map, with the *Tabula Peutingeriana* descending from a later re-edition of the same work, both being connected with supposed moments of reorganisation of the *cursus publicus* and/or the supposedly associated *annona*.[15]

Continuing a recent trend in diagnosing a distinctively Roman conception of space as fundamentally 'hodological' (perpetuating the stereotype of theoretical Greeks versus practical Romans), Kai Brodersen has recently argued that this monument was nothing but an unillustrated collection of *commentarii* on the regions of the inhabited world.[16] However, Brodersen's scepticism as to the visual quality of the Porticus Vipsania map is unwarranted on a number of counts. First of all there is the general consideration of the venerable tradition associating pictorial illustration with public porticoes, going back at least as far as the Stoa Poikile ('painted colonnade') built on the Athenian agora in the mid-fifth century BC. More specifically, the will of the peripatetic philosopher Theophrastus (*c*. 287/286 BC) stipulated that panels depicting the orbit of the earth (*períodos gês*) be set up in the Lower Stoa.[17] Second, not only the wording of Pliny the Elder's references to the Porticus Vipsania but also the two pieces of testimony that seem to describe later versions of monuments of the same genre clearly indicate that both picture and text were involved: Eumenius' panegyric *pro instaurandis scholis* at Augustodunum (Autun) of AD 298, and the epigram on the redrafting, at the instruction of the emperor Theodosius II in AD 435, of the *opus . . . , quo summa mundi tenetur* ('which comprises the entirety of the world').[18] Pliny's (and other hypothesised) quotations of Agrippa's *commentarii* suggest that the texts were largely made up of statistics concerning the overall dimensions (*latitudo*, *longitudo*) of provinces, seas, gulfs, and so on, and perhaps also the lengths of rivers. And, although major cities were clearly depicted, locating their positions within regions apparently depended on visual reference rather than the reading off of distances.[19] Furthermore, Eumenius stresses the utility of his proposed map as a visual aid to instilling geographical knowledge in the young, and asserts that his addressee has witnessed this teaching method elsewhere (*ut ipse uidisti, credo*). On the other hand, it seems unlikely that roads were depicted in maps of this type.[20] For Eumenius and the Theodosian epigram are unanimous in characterising the content of their subject as seas, mountains, rivers, harbours, straits and cities. Their silence on roads is significant. Given their visibility on the Peutinger Table, if roads had been depicted on these maps, they would surely have formed a feature worthy of remark.

While the Porticus Vipsania map and its kin cannot serve as plausible archetypes for the lists in the Antonine Itinerary or the routes on the Peutinger Table and general public knowledge thereof, a new claimant has entered the fray. There appeared in 1999 the provisional publication of an unfinished papyrus roll from Antaeopolis in Egypt that bears the first five columns of an aborted copy of Book 2 (on the Spanish peninsula) of Artemidorus of Ephesus' *Geographoúmena*, a work otherwise known only from later quotations.[21] This description of the world in eleven books was composed at the end of the second century BC, though this particular manuscript belongs half a century later. But more exciting for students of ancient cartography is the fact that this copy was clearly intended to be illustrated with maps interposed between the columns of text, since a start had been made on the first of these (in a rectangular space left between columns III and IV) before the project was abandoned and the papyrus became a later artist's sketch-pad. The map, presumably meant to depict Spain or one of its regions, contains a large number of, sometimes branching, wiggling lines running east–west, along one of which is a series of tiny squares. In addition, between the lines are to be found vignettes of individual buildings or groups within walled circuits. These are drawn with the same kind of perspective as the larger vignettes on the Peutinger Table (see Figures 3.3 and 3.4); hence the claim that this might be an earlier example of the genre, even though we are here dealing with separate regional maps on a far larger scale. Clearly the vignettes are legitimately interpreted as representing temples, forts, settlements, and so on. And while some of the wiggling lines might be rivers, it seems less plausible that the others represent routes, and in the absence of labelling it seems rash to suggest that the boxes might be the '*mansiones, Militär- und Poststationen*' of the Roman state's infrastructure.[22] If anything, to judge from the published photograph, the positioning of the vignettes in relation to the lines suggests that those that do not represent rivers might represent hill or mountain ranges. In the present state of its publication, it seems premature to hail the map of the Artemidorus papyrus as a forerunner of the Peutinger Table, certainly as far as the featuring of itineraries is concerned.

The strongest evidence against the notion that the Peutinger Table's diagram of routes and stations originates with the geographical tradition comes from the detail of the placenames themselves. For, as the following example of the route around the Bay of Naples demonstrates, the toponyms are clearly articulated as part of a sentence rather than acting as labels on a map; that is, they are in oblique cases rather than in the nominative: 'Neapoli–xi–Herculanium–ui–Oplontis–iii–Pompeis–xii–Nuceria–uiii–Salerno' (VI 4–5).[23] The same pattern is reproduced throughout the document. Thus, as far as the basis of its route network is concerned, the Peutinger Table belongs firmly to the itinerary tradition. It is a product of the realisation of the potential latent within a collection of written itineraries, such as the Antonine Itinerary, for diagrammatic representation. The network is generated by the identification of the nodal points as those that appear in more than one itinerary list, not just as beginning and end but also as intermediate stages. The Antonine Itineraries was recently subjected to exactly the same process by Ray Laurence in demonstrating the significance of Capua, Terracina, Beneventum and Aequum Tuticum as nodal points in the route network of southern Italy.[24] The nodes thus established and the intervening stages filled in, the result is a diagram that is topologically correct but topographically inaccurate; the frequently cited modern analogy being the schematic diagram of a public transport network, such as the London Underground map.[25] But was the originator of the archetype of our *Tabula Peutingeriana* alone in making this conceptual leap or but one of many? In other words, were diagrams like this commonly available to the travelling public or does the surviving *tabula* represent a unique artefact, produced perhaps more as an object of display rather than of practical utility?[26]

Certainly, that the idea of the pictorial representation of an itinerary had been heard of, at least by the late fourth century AD, is demonstrated by an oft-quoted passage from Vegetius' treatise of advice *de re militari* (3.6), even if the author claims no direct experience. For Vegetius advocates the wisdom for a commander of having '*itineraria* written out (*perscripta*), so that he might learn not only (*non solum*)' the usual information on distances 'but also about the condition of the roads (*sed etiam uiarum qualitate*)', and also so that, having had them

> accurately described (*ad fidem descripta*), he might take into account shortcuts, branch-roads (*deuerticula*), hills, and rivers. So much so, that more ingenious commanders are claimed (*firmentur*) to have had itineraries of the areas in which their attention was required not so much annotated (*non tantum adnotata*) but even illustrated (*sed etiam picta*), so that the road for setting out on might be chosen not only by a mental consideration (*non solum consilio mentis*) but truly at a glance of the eyes (*uerum aspectu oculorum*).[27]

By no stretch of the imagination can the *Tabula Peutingeriana* be said to fulfil Vegetius' desire that the physical geography be '*ad fidem descripta*', so that it seems unlikely that it could be described as an *itinerarium pictum* in Vegetius' terms. Nevertheless, it does represent a significant advance in sophistication on a collection of written itineraries, by presenting the same data in far more efficient and flexible a form. For while it lacks the already totalled distances between main towns, gone are the repetitions of routes featuring in separate itineraries. Instead, such sections are multivalent, being read easily in either direction as part of a variety of potential routes that are indeed now much more easily discerned 'at a glance of the eyes'. All this is squeezed into a roll which takes up no more space than the less 'user-friendly' manuscripts of the Antonine Itinerary. Despite these advantages the evidence of the transmission of the Peutinger Table strongly suggests that our sole example comes from a single tradition, rather than being the unique survival of a once widely represented and more venerable genre. On the internal evidence of both the depiction of Byzantium in a nimbus as Constantinople and the appearance of St Peter's at the first milestone on the via Triumphalis, the archetype cannot have been much earlier than the mid-fourth century. As for the extant artefact, although its exact provenance is unknown, its script is identified as Bavarian-Alemannic of the late twelfth or early thirteenth century, suggesting that it had not travelled far when acquired by Conrad Celtes at Worms in the last decade of the fifteenth century. It is possible that an earlier example (its archetype?) is mentioned in a ninth-century catalogue of the Abbey of Reichenau on Lake Constance and the copying of a later sister by a monk Conrad appears to be recorded in the annals of the Dominican monastery at Colmar in 1265.[28] Moreover, in our earliest witness to the family – the Anonymus Ravennas – beyond the Italian section it is only toponyms in the middle and upper Rhine valley that have received any updating.[29] It may also be significant that the only forests to be depicted and named on the *tabula*, as we have it, relate to this very region: the *silua Vosagus* (Vosges) and *silua Marciana* (Schwarzwald). Thus, whatever its ingenuity, the archetype of the *tabula* would appear to have had quite a narrow circulation.

ITINERARIES AND TRAVEL

Even without the extra sophistication offered by a route diagram, a straightforward non-pictorial *itinerarium* was still of great utility to the traveller, both in assisting in the selection of the best route and for pacing oneself once on the road. Some of the routes of the Antonine Itinerary actually advertise themselves as being a shortcut (*compendium*): for example, in Cappadocia, *A Sebastia Cocuso per Melitenam m.p. CCXCIIIS . . . Item a Sebastia Cocuso per Caesaream m.p. CCLVIII . . . Item a Sebastia Cocuso per conpendium m.p. CCVI* (176, 3–4; 178, 6–7; 180, 6–7). As for pacing, in discussion of the setting of *vadimonium* jurists reveal that the praetor allowed a speed of twenty miles per day in calculating a reasonable date on which litigants could be expected to appear.[30] The fairness of this minimum is borne out by the evidence of Theophanes' journey from Nikiu in the Delta to Syrian Antioch by means of thirty-three named stages in twenty-five days. The mean daily distance was thirty-three miles per day, the shortest achieved still twenty-four. Clearly the actual distance that could be covered in a day would be constrained by the nature of the party travelling, the contingencies of weather and road conditions, and the number of daylight hours available. If travelling along a public road, progress could be monitored by counting off the examples of that quintessentially Roman artefact, the *miliarium* (milestone, Figure 3.2), as they were passed; a fact celebrated as boon to the tired traveller by Rutilius Namatianus in his journey poem, *de reditu suo*, of AD 417.[31] Typically columnar, milestones habitually bear, apart from a commemorative text, only a single figure, which represented the distance either to or from the terminal point of the road (*caput uiae*); occasionally the distance from both ends was given.[32] Nevertheless, armed with the itinerary's list of distances between stations, the milestone's information enabled the traveller to judge the most appropriate stop each night (*mansio*, Gk. *moné*) and to calculate the number of days' travel necessary to reach

Figure 3.2 Milestone indicating 79th mile from Benerentum to Brundisium (from Keppie, 2000. Courtesy of the Egypt Exploration Society.)

his desired destination. Accordingly, Theophanes entitled his itinerary *to katamonês* (*P.Ryl.* 627 verso col. i 223).

In this way *mansio* not only came to designate the physical place of lodging but also became a way of counting the varying distance in between rest-stops. Thus, as well as absolute physical distance, a specific long-distance journey might conveniently be described in these variable units of travel, as a measurement that more directly conveyed the traveller's actual experience.[33] Accordingly, at the end of each significant section of the journey, the Anonymus

Burdigalensis gives a summary not only of distance covered and changes of animals made (*mutationes*) but also of *mansiones* taken – a total made up of overnight stops made in cities as well as in purpose-built establishments – for example, *Fit a Sirmium, Serdica usque milia CCCXIIII, mutationes XXIIII, mansiones XIII* (567, 2–3), where the last figure represents the sum of the seven *mansiones* and six *ciuitates* stayed in since Aquileia (563, 8–567, 1). This reflects a speed of travel of just over twenty-four miles per *mansio*.

Even if one's choice of routes was restricted to a single itinerary and even if one was using the facilities of the *cursus publicus*, which maintained a net-work of publicly funded lodgings (*mansiones*) and changes of animals, the data provided by an *itinerarium* on the spacing and status of possible stopping-points was still important in planning one's journey. For the likely availability of bathing facilities, shopping opportunities, and other amenities sufficient to meet the needs of the typical retinue of an independent traveller was a significant consideration in selecting one's stop-overs.[34] Such is clear from the decision-making of the imaginary commander in Ambrose's sermon in verse 33 of Psalm 118 (= 119 of the King James' Bible) and is suggested by the opportunities taken by Theophanes – who seems to have been using the *cursus publicus* – to buy bathing supplies at an early stage (Athribis), to enjoy public entertainments (at Ascalon), and, on the return journey, to partake of the exotic delicacy of 'snow-water' (at Biblos).[35] The possibility of being able to stop in a decently sized settlement for lunch also seems to have been an important consideration, at least to judge by Theophanes' pattern of behaviour.[36] Given such considerations, in cases of a choice between routes, the shortest was not always the best, especially if it meant traversing a less densely populated region, with all the discomforts that might entail. This, rather than a fundamentally different perception of geography, was, I believe, the major factor (with the not unconnected fact of the relative ease of the terrain) in determining the greater popularity of the longer of the two land routes from Gaul to Rome: that is via the Po valley and then across the Appennines by the via Flaminia, rather than along the Ligurian and Etruscan coast.[37] This example and the case of Theophanes, who descended from Hermopolis to the Delta by the Nile, reminds us that in many instances a journey was best made in whole or part by water. For from Rome to Gaul the easiest journey would be by coaster from the Tiber to the Rhône, the very journey commemorated by Namatianus' *de reditu suo*. Thus an itinerary that comprises terrestrial routes alone represents only one side of the pattern of travel as actually experienced by Romans. The point is illustrated by the third *Satire* of Lucilius Iunior, which describes a journey from Rome to Sicily. The poet considers the distances with care, and then opts to take the via Appia to Campania but at Puteoli to switch to marine transport. Thence he coasts, via Salernum and Paestum, to Cape Palinurus in Lucania. At that point he takes to the open sea, heading for the Straits via the volcanic Aeolian islands.[38]

While Theophanes wrote his return itinerary along the same route out again (interestingly including at least one place at which he did not call on the way out),[39] this was not, of course, necessary; as is demonstrated in the other surviving itinerary explicitly relating to a specific journey travelled – that of the Bordeaux pilgrim(s). This bears the hallmarks of being a neat copy of the kind of document represented by Theophanes' travel memoranda. The careful categorisation of each location as *ciuitas*, *mansio* or *mutatio* strongly suggests that the pilgrims were also making use of the facilities of the *cursus publicus*, whether at the state's expense or their own. Most notably, the Palestinian section is filled out with extensive notes on visiting sites that feature in the Old and New Testaments, for which the work was much prized in the Middle Ages.[40] On a more mundane level, although the minutiae of expenditure found in Theophanes' rough notes are absent, the drafter of the Bordeaux Itinerary does appear to have integrated observations made on the places passed.[41] The vast majority of these scholia are impersonal but in six places they are expressed in the second-person singular ('you cross', 'you enter', 'you travel', etc.).[42] However, if copied up with regard to a future user, and assuming that its writer did make the entire return journey to Bordeaux, the text does not represent an exhaustive record, as the manuscript title indicates: *Itinerarium a Burdigala Hierusalem usque, et ab Heraclea per Aulonam et per urbem Romam Mediolanum usque* (549, 1–5). To be precise, the itinerary runs continuously from Bordeaux to Constantinople, via Milan, Aquileia, Sirmium (modern Sremska Mitrovica), Singidunum (Belgrade), Serdica (Sofia), Philippopolis (Plovdiv) and Heraclea (Ereǧli) on the Sea of Marmara (549, 7–571, 5). After a round-up of the totals of stages and distances thus far, the itinerary starts afresh, giving a date for departure, from the Bosphorus to Jerusalem, via Ancyra (Ankara), Tarsus, Antioch, Tyre and Caesarea (571, 9–589, 6). After using Jerusalem as the base for tourism in the region, a different return itinerary is outlined to Caesarea, via the coast (600, 1–601, 5). There follows a huge jump to Thrace, since the itinerary resumes with the route from Heraclea to Milan. However, this time it traverses Macedonia and, via the crossing from Aulon (Vlorë) to Hydruntum (Otranto), continues up the via Traiana and Appia to Rome, before regaining Milan by the via Flaminia and Aemilia (601, 6–617, 8). In this way the entire round trip is accounted for in the most efficient manner; the missing links (Caesarea–Heracleam and Mediolano–Burdigalam) can be filled in by tracing the appropriate sections of the outward itinerary in reverse order because the disjointed sections of the return itineraries have their starting and terminal points at locations which featured in that outward itinerary. This very practice is indicated by the one first-person comment, inserted in the outward itinerary at the Bosphorus (571, 6–8): *Item ambulauimus Dalmati{c}o et Zenophilo cons. III kal. Iun. a Calcedonia* [sc. *uersum Hierusalem*], *et reuersimus Constantinopolim VII kal. Ian. cons. suprascripto* (i.e. 'Similarly, we travelled from Chalcedon on 30 May AD 333, and came back in the other direction to Constantinople on 26 December of the same year').

ITINERARIES AND ARCHIVES

Both the itinerary of Theophanes and that of the Bordeaux pilgrim have been tailored to the needs of a particular journey. What might have been the source from which their itineraries were acquired? As noted above, the facilities of the *cursus publicus* seem to have been at the travellers' disposal in both cases.[43] From the wording of letters of introduction to provincial governors included in his papers, it is clear that Theophanes was travelling on the pretext of official business.[44] Moreover, there is room for doubt as to whether the journey from Bordeaux was intended as a Holy Land pilgrimage from the outset. Despite what the manuscript title says, strictly speaking the trip from Bordeaux to Constantinople is one discrete journey. The section onwards to Jerusalem is then signalled as a separate journey proceeding from Chalcedon (571, 6–8, quoted above). Another circumstantial indication is that one might expect a serious pilgrim to have planned the trip in order to have been in Jerusalem at Easter.[45] It is possible, then, giving full weight to the first-person remarks, to imagine that a single individual (the 'narrator') set out from Bordeaux, travelling by the *cursus publicus*, in pursuit of some kind of official business in Constantinople (it may be significant that the subscriptions of five imperial laws indicate that the court of Constantine was resident there between 17 October 332 and 5 May 333).[46] Business there concluded, and having successfully elicited a permit to extend the use of the *cursus publicus* to take in the opportunity to visit Palestine, he or she took up with a party leaving Chalcedon on 30 May. The group then set off on their tour to Jerusalem and its environs, before returning to Constantinople on 26 December, where they might have parted company once more, the narrator to make a slightly more leisurely return to Bordeaux. If this reconstruction is correct, then the Bordeaux 'pilgrim' is revealed as first and foremost another traveller on official business.

The conclusion that the journeys of both Theophanes and the Anonymus Burdigalensis were prompted by official business seems to offer support to the persistent idea that, even if not associated with a pictorial diagram, our manuscript itineraries might ultimately derive from documents issued by the administration of the *cursus publicus* and/or the *annona*, whether in connection with the planning of imperial journeys, the movement of troops or the travels of individual soldiers.[47] However, as Oswald Dilke pointed out,[48] the evidential basis for this view is small in size and dubious in nature, resting as it does on two late fourth-century literary texts. The first, a passage of the *Historia Augusta*'s life of Severus Alexander, supposedly bears witness to a practice of the imperial court, whereby the emperor's travel plans were published in advance in order to allow arrangements for lodging and provisioning to be prepared:

> The secrets of campaigns were kept silent but the dates of journeys (*itinerum dies*) were publicly proclaimed, so that he [Alexander] would

> even issue the edict (*edictum penderet*) two months in advance, in which was written: 'On that day, at that hour I shall be setting out from the city and, if the gods are willing, I shall be staying at the first *mansio*,' and thereafter in order (*per ordinem*) the *mansiones*, then the stopping-places (*statiuae*), then where the *annona* was to be collected, and thus it might indeed go on all the way up to the barbarian frontiers.
>
> (SHA, *Sev.Alex.* 45. 2)

The second, the aforementioned sermon of Ambrose, has been claimed to demonstrate that soldiers were regularly issued with a travel document before setting out on military business:[49]

> Let us listen, I say. What will he [the psalmist] have had to say about how he would live?: 'Lay down the law for me, O Lord, the way of Thy righteousness; and I shall always seek it out.' The soldier who is setting out on a journey, does not organise his own travel arrangements (*uiandi ordinem*), nor does he choose his route (*uiam carpit*) by his own judgement, nor chase after the pleasurable shortcuts (*uoluptaria captat compendia*), lest he slip away from the standards (*ne recedat a signis*); rather he takes his itinerary from the commander (*itinerarium ab imperatore accipit*), and he keeps to that: he proceeds according to the prescribed stages (*praescripto incedit ordine*), travelling with his weapons, and makes the journey by the right route (*rectaque uia conficit iter*), so that he will find the assistance for his journeyings already prepared. If he travels another route, he receives no *annona*, and does not find the *mansio* readied for him. Given that the commander orders that everything be prepared for them, those who follow (*qui secuntur*) do not stray to the right or left from the prescribed path (*a praescripto itinere*). He, who follows his commander, is not lacking in merit; for he travels in a steady manner, since the commander considers not what is to his own advantage but what is achievable by everybody: for that reason he establishes rest-stops (*statiuas ordinat*); an army travels for three days, on the fourth day it rests. Towns are chosen, in which three, four, or more days might be spent, if they abound in waters and are bustling with trade; and thus the journey is made without effort, until that city – which is chosen as if almost heavenly – is reached, in which rest might be supplied to the tired armies.
>
> (Ambrose, *Expositio psalmi* CXVIII 5.2)

Certainly the verbal echoes between the two texts are initially persuasive of the idea that written itineraries, laying out the progress of official parties in association with the preparation of provisions for them, were issued as a regular practice. This is founded on an elision of Alexander's *edictum* in the *Historia Augusta* with Ambrose's soldier's *itinerarium*. Accordingly the manuscript itineraries are taken to represent the preservation of such travel documents as issued or the citation of potential itinerary documents from government files. However, before this can be admitted, greater consideration must be given to the context of both passages.

It is certainly the case that the movements of the imperial court were a considerable logistic problem that might be a strain on the resources of the area through which it passed. Cassius Dio complains of the burden put on the local land-owning class by Caracalla's profligate attitude to travelling, particularly by the wasted effort expended in making preparations in anticipation of visits the emperor never ultimately made (Dio 78.9.3, 6–7). The impression gained is that the court gave out piecemeal and sometimes contradictory information. It is not a surprise that the behaviour of the *Historia Augusta*'s Alexander should be in such contrast. For the ideological agenda is particularly clear in the account of his reign, which forms with the preceding biography of Elagabalus a focal section of the work. The awful Elagabalus is contrasted with the virtuous Alexander, whose rule is in many ways presented as that of the paragon of the good emperor. Thus, appearing as part of a eulogy on the open character of his government, Alexander's supposed publication of his intentions two months in advance surely represents the wishful thinking of the author. As such, it can scarcely be taken as fact. As for the itinerary of Ambrose's humble soldier, it must be borne in mind that the bishop surely intended his Milanese congregation to understand the soldier travelling according to his commander's orders as an analogy for the faithful Christian living his life according to the word of God. The physical reality of the *itinerarium* received by the soldier is a construction of the reader not a necessity of the text. However, it is this presupposition that has enticed Brodersen, for instance, into interpreting – unnecessarily from a grammatical point of view – Ambrose's *qui secuntur* ('those who follow') in a quasi-transitive manner, with *itinerarium* as the understood object.[50] If an understood object for *secuntur* had to be sought, the concrete *eum* (sc. *imperatorem*), rather than the abstract *itinerarium*, would be much more natural in the context (i.e. 'those who follow [their commander] do not stray to right or left'). As a general consideration, the soldier's possession of a written itinerary might make sense if he were travelling alone; but it is clear from the analogy as Ambrose develops it that the bishop imagined his soldier as travelling in the commander's train.

If, then, the external evidence for the link between the dissemination of itineraries and the administration of the *cursus publicus* or *annona* is not forthcoming, what of internal evidence. The most powerful evidence would seem to be the consistent mention of locations as *mansiones* and *mutationes*, that is, the paraphernalia of the imperial postal system, throughout the Bordeaux Itinerary, for example (556, 5–10):

inde incipit Italia
mutatio ad Duodecimum *milia XII*

mansio ad Fines	*milia XII*
mutatio ad Octauum	*milia VIII*
ciuitas Taurinis	*milia VIII*
mutatio ad Decimum	*milia X*
mansio Quadratis	*milia XII*
mutatio Ceste	*milia XI*
mansio Rigomago	*milia VIII*
mutatio ad Medias	*milia X*
mansio Laumello	*milia XII*
mutatio Duriis	*milia VIIII*
ciuitas Ticino	*milia XII*
mutatio ad Decimum	*milia X*
ciuitas Mediolanum	*milia X*

However, as this extract makes equally clear, the categorisation of the locations, which is consistent in form throughout the manuscripts, stands outside the grammar of the itinerary list itself, being in the nominative in contrast to the oblique cases of the locations. Just as with the lengthier descriptions of Palestine, these would appear to derive from a gloss on the itinerary document made by its user, describing the nature of their progress: 'a change', 'a stay', 'a city', and so on. The underlying itinerary is thus no different in form from those found in the Antonine Itinerary or Peutinger Table. Thus the Bordeaux Itinerary may well attest to the use of the facilities of the *cursus publicus* but it cannot serve as unquestionable evidence for that organisation as the source of its list of places. Nor does the codicology of the Antonine Itinerary suggest such an official origin by observation of the company it keeps. It is true that one copy was contained in the now lost Carolingian *Codex Spirensis*, along with the *Notitia Dignitatum*, a late Roman work of a technical nature with an undoubtedly official origin. However, this is an isolated case and the *itinerarium* is consistently found in other manuscripts, earlier and later, in the less exalted company of Iulius Honorius' *Chorographia*.[51] What then might legitimately be hypothesised about the source(s) of the itineraries in the Antonine Itinerary and Peutinger Table on the basis of their internal make-up?

THE STRUCTURE OF THE ITINERARY COLLECTIONS

The Antonine Itinerary

This comprises a series of itineraries touching upon all regions of the Roman empire as it appeared in the last quarter of the third century AD; that is, it omits any coverage of routes across the Agri Decumates in the angle between the Rhine and Danube or in the transdanubian province of Dacia, both areas which were abandoned in the 270s. Indeed, while the information in some sections probably goes back as far as the Severan period, there are numerous signs of tetrarchic material, not least the presence of Diocletianopolis in Macedonia (330, 6). On the other hand, that Byzantium is qualified only once as '*qui et Constantinopoli*' (138, 5), and then not in the earliest manuscript, suggests a date for the collection before the city was rebaptised in AD 324. Indeed, there is nowhere any sign of the Christianisation of the toponymy

The Antonine Itineraries' survey is organised roughly into regional sub-groups, sometimes explicitly so titled.[52] The text begins, as was common in ancient geographical writing, from the Pillars of Hercules (actually illustrated and so labelled in the earliest manuscript, itself of Spanish origin).[53] However, somewhat less canonically, it proceeds in a counter-clockwise direction around the shores of the Mediterranean from Mauretania to Africa and, having listed routes through Libya to Egypt, skips across to the Tyrrhenian islands (Sardinia, Corsica and Sicily) and on to Italy. A single itinerary from Rome to Hiera Sycaminos then takes the reader through the Danubian provinces, Anatolia and Syria, to the southern reaches of Egypt (123, 8–162, 4). The description then doubles back to fill out the coverage of other routes in the provinces traversed in the previous itinerary, as well as the Alpine regions, before re-entering Italy. Routes the length of the peninsula are then described, leading, via a crossing of the Adriatic to Epirus, to a description of routes in Dalmatia, Achaea and Macedonia, even including the northern part of Asia by means of a crossing from Thrace to the Troad. There then follows a distinct break, since the focus moves to routes from Milan into Gaul, then to a broad coverage of Spain, which section is concluded by recrossing the Pyrenees to a description of routes in Aquitania. The provincial itineraries are rounded off with a section devoted to Britain. Thus, rather than a tidy circular survey of the routes of the empire, the trajectory of this collection is closer to a figure-of-eight.

As is obvious at a casual glance, the structure of the itinerary description is not uniform. Some sections are dominated by the explication of a single long journey that forms the spine of the section, to which the description of shorter more local routes are pegged. In a couple of cases the apparent eccentricity of the route taken indicates that the planning or recording of a specific real journey is reflected, in which priorities other than simply taking the quickest and most convenient path from start to finish clearly dictated the route. The most famous example is the itinerary from Rome to Hiera Sycaminos, which is generally considered to form the backbone of the entire work. Those who have upheld the manuscript attribution to an emperor Antoninus wished to see this as the original core document relating to a trip by Caracalla in 214–15. However Pascal Arnaud has demonstrated that even here there lurk elements of indubitably tetrarchic date.[54] Another notable example is the '*Iter quod a Mediolano per Picenum et Campaniam ad Columnam, id est traiectum Siciliae, ducit*' (98, 2–106, 4). This does not take what would seem the easiest route: taking the via Aemilia and Flaminia to Rome, thence on down the via Appia to Capua, and on via Salernum and Cosentia to the Straits. Rather, at Fanum Fortunae, instead of following the via Flaminia to Rome, it takes a far more arduous route, continuing down the Adriatic coast to a *uicus* on the River Aternus (i.e., Pescara), where it turns inland to Corfinium and thence southwards along the Appennine ridge to Venusia, before cutting across to Nerulum to join what is acknowledged as the more conventional 'direct route' from Rome to the Straits (*ab urbe recto itinere ad Columnam*: 106, 5).

More frequently, the text comprises clusters of routes listed to and from a nodal point, such as Rome (289, 3–315, 6), Milan (339, 7–358, 4) or London (473, 1–480, 8); a pattern that is also seen in the structuring of Pausanias' guide to Greece.[55] Moreover, as emphasised by Arnaud, in particular stretches the description of the places exhibits a distinctive style. For example, a significant number of the African itineraries note the political status of the settlements along the route as *colonia*, *municipium*, *ciuitas* or simple *uicus* (e.g. 12, 1–19, 2; 36, 3–41, 2; 58, 1–63, 2) and two routes that have the common feature of a traversal of northern Italy concern themselves with the slightly simpler question of whether each place is a *ciuitas*, *mansio* or *uicus* (98, 6–102, 5; 124, 8–132, 1). These Arnaud sees as the certifiable relics of earlier itineraries dismembered and then reassembled according to the compiler's grand scheme. It may also be true that these sections derive from an itinerary document that has been glossed through use in the same way as the Bordeaux Itinerary.

While much of the regional coverage is interlinked by itineraries joining one province to another and often labelled as such,[56] there are a number of places in which there are distinct discontinuities. Sometimes these are forced by the geography (as in the description of islands), but more commonly not. Rather they might be taken to reflect the boundaries between regional sub-collections. By observing such discontinuities, up to ten separate sub-collections may be identified:

I. Mauretania–Numidia–Africa–Tripolitania–Cyrenaica–Aegyptus (1, 1–78, 3)
II. Sardinia (78, 4–85, 3)
III. Corsica (85, 4–86, 1)
IV. Sicilia (86, 2–98, 1)
V. Italia–Noricum–Pannonia–Moesia–Thracia–Bithynia–Galatia–Cappadocia–Syria–Palaestina–Aegyptus–Cyrenaica (98, 2–173, 4)
VI. Thracia (175, 1–176, 2)
VII. Cappadocia–Syria (176, 3–217, 4)
VIII. Moesia–Thracia–Bithynia (217, 5–231, 8)
IX. Pannonia–Noricum–Raetia–Gallia, Pannonia–Dalmatia–Italia–Raetia–Noricum, Italia–Raetia–Noricum, Italia–Epirus–Macedonia–Thracia, Epirus–Macedonia–Thracia–Asia, Dalmatia–Epirus (231, 8–339, 5)
X. Italia–Gallia–Hispania–Aquitania–Gallia, Britannia (339, 6–486, 17)

It would seem that these have been stitched together by a compiler to produce the overall provincial itinerary. Significantly the breaks between these conglomerations are not entirely coincident with the rubrics, which were presumably imposed by the 'author', or rather compiler, of the whole work. Not only have logical divisions within these sections been highlighted but also, conversely, the breaks separating the sixth, seventh and eighth sections (176, 2–3 and 217, 4–5) are marked by no special titles at all to alert the reader, despite the quite dramatic jumps (in the first instance from Heraclea on the Sea of Marmara to Sebastea in Cappadocia, in the second from Satala on the Armenian border to Viminacium on the middle Danube). This would suggest these sections represent a pre-existing mini-collection of itineraries produced by an earlier compiler whose practice was not as reader-friendly. Within these mini-collections can be found material of disparate pattern, according to Arnaud's diagnosis, as well as of disparate date. In the sections dealing with the provinces of Gallia Comata there not only exist the original names for the *civitas*-capitals alongside the ethnic equivalents that were in widespread use by the fourth century, but also itineraries still genuinely measured (as opposed to mistakenly labelled) in miles rather than leagues, the unit of all-new road-surveying in this region from *c*. AD 200.[57]

The accompanying collection of maritime itineraries is similarly variegated and disjointed. Moreover, from the very beginning it fails to live up to its claimed intention, that it will instruct the reader which shores to follow and where to sail starting from the Pillars of Hercules (*incipiens a Gadibus uel extrema Africa*: 487, 4). For it immediately proceeds to describe a course from the Isthmus (of Corinth) via Sicily to Carthage. There follows a list of crossings (*traiectus*) between various points in the western Mediterranean, including, it is true, that from Baelo (modern Bolonia in Andalusia) to Tingis (Tangier) but the itinerary from Carthage to Tangier is nowhere filled in. This deficiency could in part be made good by the acute reader of the collection, if they remembered that the very first section of the provincial itinerary incorporated a journey by sea along the coast from Tangier as far as the Portus Divini in Mauretania Caesariensis (*A Tingi litoribus nauigatur usque ad Portus Diuinos*: 9, 1–13, 7); though this is still some 894 Roman miles short of Carthage, according to the same itinerary. The maritime itinerary continues by listing further *traiectus* – one across the English Channel and several in the Adriatic – after which there is a sharp break, marked by the listing of the harbours and anchorages ('*itinerarium portuum uel nauium positionum*': 497, 9–10) along the coast from Rome to Arles on the Rhône in southern Gaul. Until this point the maritime itinerary had listed distances in stades, following the Greek seafaring tradition from whose *períploi* its information no doubt derived. However, as with the earlier itinerary along the Mauretanian coast and others around Sicily also in the provincial section (90, 6–93, 1; 95, 2–96, 4; 97, 7–98, 1), the distances are measured in Roman *m(ilia) p(assuum)*. This peculiarity puts these coastal itineraries in the same tradition as the enigmatic document from Syria known as the 'Dura Shield' (Figure 2.2). Dating from before that city's capture by the Persians in AD 260, this scrap of parchment (roughly 18 × 6.5 cm), of which the centre is decorated with ships filled with sailors, similarly lists the distances in miles between cities and features along the Black Sea coast around its edge, even though the lettering is Greek.[58]

After reaching Arles the maritime itinerary changes tack again, launching into a list of islands, beginning with those around Britain (no distances) and continuing with those between (*inter*) various regions in the western Mediterranean and Adriatic, with distances from the shore once again in stades. After this, the collection peters out with a disorganised and somewhat confused listing of islands in the Aegean and eastern Mediterranean, to a reasonable number of which are attached notes illustrating their role in classical mythology. There is no sign of the impact of Christianity here. Patmos is not even listed, let alone associated with the author of the Apocalypse. This strongly suggests the excerption of this material from one of that genre of works that illustrate the geography of Homeric epic.[59] In sum, the maritime itinerary illustrates in a more extreme form the pattern clearly visible from the diagram of the *itineraria* shown in Figure 3.1: that is, the relative paucity of the Antonine Itinerary's coverage east of Pannonia and Tripolitania. Though, conversely, the land routes also thin out considerably to the east of the Loire–Rhône axis, so that it would be more accurate to say that it is the central tranche of the empire that receives the fullest coverage, with Milan as one of the most significant nodes. The almost total lack of coverage for Asia Minor, the Aegean and Old Greece, compared with the density of routes for such areas as Pannonia and the Armenian border, rules out any connection with aristocratic cultural tourism.[60] As a whole the pattern of the data in the collection suggests a western-based utilitarian viewpoint, while the presentation reflects a desire to appear as comprehensive as possible in geographical coverage.

The Peutinger Table

This displays many points of contact with, as well as of difference from, the Antonine Itinerary. Its route network has clearly been drawn up from a similar kind of itinerary collection. On land its coverage is both more ambitious (taking in India and the Far East) and more comprehensive (having fewer gaps inside the empire), as well as being more detailed, though neither is any more trustworthy than the other in the correlation of its distance figures with reality. For example, for the route from Arretium to Rome:

Tab. Peut. IV 4–V 5		*It. Ant. 285, 4–286, 5*	
Adretio		*Arretio*	
ad Nouas	*xxiiii*	*ad Statuas*	*m.p. XXV*
Clusio	*uiiii*	*Clusio*	*m.p. XXV*
Pallia fl.	*uiiii*		

Volsinis	*<xxi?>*	*Vulsinis*	*m.p. XXX*
Aquas Passaris	*uiiii*		
Foro Cassi	*xi*	*Foro Cassi*	*m.p. XXVIII*
Vico Matrini	*iiii*		
Sutrio	*<uii?>*	*Sutrio*	*m.p. XI*
Vacanas	*xii*	*Baccanas*	*m.p. XI*
Veios	*uiiii*		
ad Sextum	*ui*		
ad ponte(m)			
Iulii {sic}	*iii*		
Roma	*iii*	*Roma*	*m.p. XXI*

The *Tabula Peutingeriana* also uses the same red ink as it does for the land network to pick out the banks of navigable rivers, at least for northern Italy and central and southern Germany, suggesting that the underlying itinerary collection included an *itinerarium fluuiale* for this area. On the other hand, open-water maritime itineraries are completely absent, though coasting sections have occasionally become inadvertently transformed into land routes: for example, that linking Miletus in Asia to Patara in Lycia by way of Myndus, Cnidus and Loryma (X 1). In terms of chronology the itineraries employed are even more disparate than those in the *Itinerarium Antonini*. As noted above, the cities destroyed by Vesuvius in AD 79 manage to coexist with the Constantinian St Peter's. In fact the Christianisation is rather haphazard. Thus, in Palestine while the Mount of Olives and the River Jordan have been rubricated, Jerusalem is labelled: *antea dicta Herusalem n(u)n(c) Helya Capitolina* (X 1). Also in discord with the fourth-century features is the inclusion of routes for trans-Danubian Dacia (VII 3–VIII 3) and the eastern half of the Agri Decumates (III 5–IV 1). This variegated nature makes the attempt to date the whole on the basis of the omission, inclusion or highlighting of any particular location a fruitless exercise.[61] Moreover, as Arnaud has observed, the chronological variety of the data reflects not the work of layers of subsequent redactors but rather the differences in the dates of the sources used by the cartographer for each region.[62] Thus in Gaul we have the same mix of traditional names for the *civitas*-capitals with the ethnic shorthand forms, and in North Africa the same predilection for recording the civic status of settlements (*municipium*, *colonia*), as found in the Antonine Itinerary. These incongruities are hard to square with a contemporary official origin for the data or state sponsorship of the project. Private initiative seems much more plausible. It seems most likely that we are dealing with the result of the initiative of one private individual, who cannot be expected to have been equally well informed and up-to-date in his information on every area of the empire. In fact, there are traces of such a personal focalisation.

The personifications of the cities, and particularly those icons of the second rank, are most telling (Figures 3.3 and 3.4). Leaving aside the programmatic

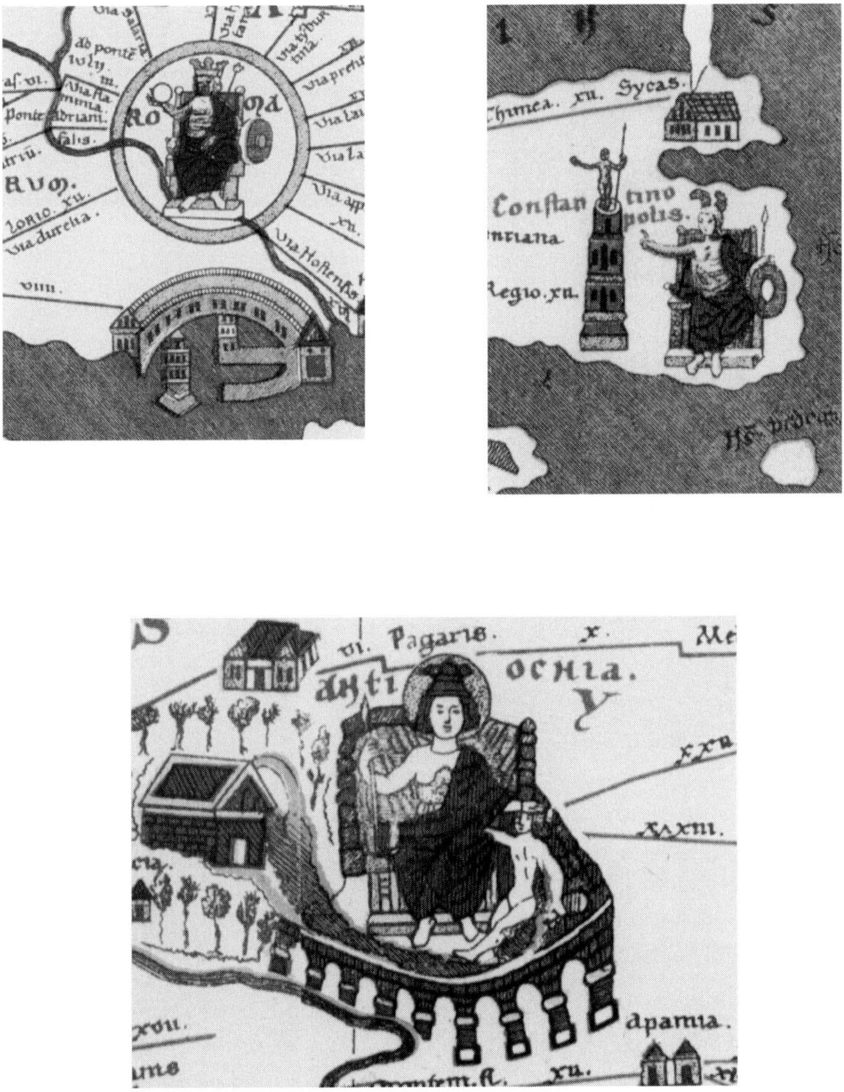

Figure 3.3 Peutinger Table: cities represented by personifications (after Miller). Clockwise from top left: Rome, Constantinople, Antioch. (With permission from the Österreichische NationalBibliothek.)

Figure 3.4 Peutinger Table: cities represented by walled circuits (after Miller). From top to bottom, left to right: Ravenna, Aguileia, Thessalonica, Nicomedia, Nicaea, Ancyra. (With permission from the Österreichische NationalBibliothek.)

placement of Rome at the centre, the two cities to be given personifications are Constantinople and Antioch. On the third rank, depicted by walled circuits with buildings within, are Ravenna, Aquileia, Thessalonica, Nicomedia, Nicaea, and Ancyra (unlabelled). Below this there is little further differentiation in depiction, with most places to get an icon being given twin towers, though locations named as bath complexes, granaries, and headquarters (*praetoria*) are given greater characterisation.[63] Almost absurdly Milan (IV 2) shares the same symbol (two turrets on a base) with Ostia Aterni, Antium (both VI 1) and the obscure station of Castra Minervae in the heel of Italy (VII 2). Nor does it form a significant node in comparison with its high profile in the *Itinerarium Antonini*. Given also the absence of Alexandria, Carthage and Trier, it is hard to construe the choice of the cities of the second and third ranks as representing any acknowledged state hierarchy. Rather, it seems to reflect a selection even more personally idiosyncratic than that of Ausonius in his *Ordo urbium nobilium*.[64] In combination with the focus of the *itinerarium fluuiale* and the *siluae*, one might plausibly interpret this as the choice of a native of north-eastern Italy who was familiar with the transalpine provinces to the north and who had also ventured as far as Antioch, not too far from a mirror version of Ammianus Marcellinus' experience of the world, reflecting the balance of imperial activities of the mid- to late fourth century rather than the traditional cultural geography of the Roman world.[65] It is reasonable to suppose that the inspiration for the creation of the diagram of routes may have derived from the anonymous compiler's intensive use of the collection of written itineraries that underlies it in the course of planning and undertaking such journeys. So the *Tabula Peutingeriana* may, after all, be a document related to a personal experience of travel.

The Antonine Itinerary and Peutinger Table would both appear to be based on disparate, but different, collections of written itineraries, incorporating at least some, analogous to the Bordeaux Itinerary, that have received glosses through use. Given the great variation in the age of the source data, even more acute in the *Tabula* than in the Antonine Itinerary, an origin in the archives of the *cursus publicus* seems unlikely. That is not to say that the administrators of the *cursus* and organisers of imperial journeys did not maintain such records; simply that these manuscript itineraries do not derive from them. Nor is it to say that the information depends on independent survey. It most certainly does rest on the data generated in the original surveying of the roads and erection of their milestones and is thus official in that sense. Thus variations in the standards used for the mile or league, which may be apparent in some of the data, plausibly originate in slightly variant standards used by the original road engineers.[66] However, the supposed external evidence disposed of, it must be acknowledged that the nature of the internal structure of the manuscript itineraries does not easily accord with the notion of official publication. It is at this point that the epigraphic evidence needs to be introduced.

TABELLARIA

For several epigraphic itineraries exist, though they have tended to be treated as parallel phenomena to the manuscript itineraries, as texts rather than monuments, and similarly as documents of geographical *Weltanschauung*. On the other hand, when their role in the practice of travel and the publication of itinerary information has been considered there has been a tendency to treat them as little more than glorified milestones or to associate them with non-itinerary artefacts, such as those souvenirs of Hadrian's Wall, the Amiens Patera and Rudge Cup; rarely has their relationship with the manuscript itineraries been explored.[67]

The most obviously attractive candidate for an itinerary stone would appear to be the *miliarium aureum*, which was erected at the head of the Roman forum near the Temple of Saturn by Augustus in connection with his rebuilding of public roads in Italy – which might also naturally be associated with the establishment of the *cursus publicus*. This 'golden milestone' Plutarch describes as the point 'at which all the roads that transect Italy terminate' (*Galba* 24), which certainly indicates a symbolic role akin to that performed by Charing Cross in the road network of modern Britain. On this basis it has been claimed as an itinerary stone detailing the stages out of Rome along the major *uiae publicae* but the only statistics it seems to have borne for certain are those quoted by Pliny the Elder in *Natural History* 3. 66: the distances from this pillar to the various gates of the city, from which points distances to places beyond were habitually measured.[68]

While it seems best to consider the case for the *miliarium aureum* not proven, there is one significantly older inscription that not only contains distance data within it but may be able to shed light on the whole issue of itinerary stones in general. This is that most enigmatic document, the so-called '*elogium*' from Polla, a village in the Vallo di Diano in Lucania, which forms the natural communications corridor between Campania and the toe of Italy.[69] Inscribed in archaic capitals on a rectangular block of limestone, 70 cm tall and 74 cm wide, this text records the achievements of a now nameless Roman magistrate, including the building of the road from Rhegium (modern Reggio di Calabria) on the Straits of Messina, through this unnamed place and on to Capua in Campania, from whence the via Appia could take the traveller on to Rome. Without the name of its author the text seems logically incomplete but the stone may well be undamaged, for it is currently to be found near the Taverna del Passo at Polla, encased in a concrete monument of the Fascist era in front of the now defunct Consorzio Agrario. Its unrecorded thickness front to back is therefore not presently ascertainable, nor is it possible to examine its top surface in order to determine whether the stone directly supported a statue, on which the magistrate's name could have been carved, or an intermediary plinth that bore this information, as some suppose.[70] It reads:

uiam fecei ab Regio ad Capuam et
in ea uia ponteis omneis miliarios
tabelariosque poseiui hince sunt
Nouceriam meilia LI Capuam XXCIIII
Muranum LXXIIII Cosentiam CXXIII
Valentiam CLXXX{{}} ad Fretum ad
statuam CCXXXI{{}} Regium CCXXXV{{II}}
suma af Capua Regium meilia CCC
et eidem praetor in XXI{{}}
Sicilia fugiteiuos Italicorum
conquaeisiuei redideique
homines DCCCCXVII eidemque
primus fecei ut de agro poplico
aratoribus cederent paastores
forum aedisque poplicas heic fecei.

(*ILLRP* 454 = *Inscr.It.* III 1, 272)

Which, without elegance but retaining the original's paragraphing, I translate as:

I made the road from Rhegium to Capua and on this road placed all the bridges and the *miliarii* and *tabelarii*. From here (there are) 51 miles to Nuceria, 84 to Capua, 74 to Muranum, 123 to Cosentia, 180 to Valentia, 231 to the Statue at the Straits, 237 to Rhegium; 321 miles in total from Capua to Rhegium.
And I, the same man, as *praetor* sought out in Sicily fugitives of the Italians and returned 917 men.
And I, the same man, was the first to see that concerning public land the herdsmen should yield to the ploughmen.
Here I built the forum and public buildings.

Despite its apparently straightforward nature, the text remains controversial in almost all respects. Its tantalising incompleteness and laconic ambiguities have made it the subject of continuing scholarly debates embracing a range of questions, including the identity of its author (and consequently the name and date of his road), in what capacity (e.g. consul or praetor) he achieved this, the name of the community which benefited from his public munificence, the relationship to the Slave Wars of his activities in rounding-up fugitives, and to the Gracchan land redistribution schemes of his encouragement of arable farming, as well as the text's apparent internal linguistic inconsistencies. None of which need unduly detain us here. Suffice it to say that the majority of the identification or dating solutions – historical, prosopographical and linguistic – tend to cluster in the third quarter of the second century BC.[71] The particular controversy which concerns us is that over the nature and definition of the *tabelarii* mentioned in the third line of the Polla text; a debate which has been simmering away now for over a century.

At first glance this word would appear to present no problem. As a noun, *tabellarius*, to give it its classical orthography, is a perfectly well-attested term in Latin prose with the usual meaning of 'letter-carrier'.[72] However, its use here paratactically alongside the *pontes* and *miliarii* strongly suggests that another item of inanimate road furniture is intended. As to what was meant precisely, there can hardly be said to be a consensus of opinion. For there are currently at least seven different interpretations of the meaning of *tabellarii* in this context. In the 1850s Theodor Mommsen had originally understood the *miliarios* to have been simple milestones, with only an indication of the number of miles, and the *tabelarios* to have been indicators carrying an inscription of more than simply the number and presented in the form of a *tabella*, that is, a tablet-stone (*Tafelstein* or *Weisungstafel*), as shown in Figure 3.2. However, by the time he came to write the commentary to the Polla inscription for the first edition of volume one of the *Corpus Inscriptionum Latinarum* he had changed his mind somewhat. He now considered the *miliarii* to be the familiar type of dressed, cylindrical milestones and the *tabellarii* to be flat (smooth) and quadrangular milestones; a type found alongside the more familiar columns up to this period but which seems to have died out thereafter.[73] On a slightly different tack, in the early years of this century Otto Hirschfeld suggested that the *tabellarii* differed from *miliarii* in that their lettering was not chiselled directly into the stone, but rather cut into a tablet of wood or bronze affixed to their surface. He argued that these materials, being easier to engrave than stone, were more suitable for the longer text.[74] While not necessarily accepting every detail of Hirschfeld's solution, the idea that the *tabellarii* represented a newly introduced type of milestone has been followed by a number of scholars, including, most recently, Brodersen.[75]

A simultaneously popular solution has been simply to understand *tabellarius* with the well-known meaning 'messenger', and hence consider the Polla text as evidence of the establishment of an otherwise unknown precursor to the public postal system of the imperial period (the *cursus publicus*). This interpretation has continued to receive constant, if minority, support up to the present, having been most recently vigorously restated by Sylvie Crogiez, anxious to utilise the Polla text in tracing the stations of the *cursus publicus* in Lucania-Bruttium.[76] However, objections, both historical and linguistic, have been raised to the translation of *tabellarii* here as 'messengers of the postal service'. First, the regular *cursus publicus* was not instituted until a century later by Augustus and there would, in any case, have been less of a need for the relatively autonomous magistrates of the republican period to maintain such a communication system with Rome. Second, it seems difficult to construe the verb (*ponere*) simultaneously with objects designating infrastructure material (*pontes*, *miliarii*) and a service assumed by people (*tabellarii*). The latter consideration led David Stockton to the compromise translation 'posting-stations' (i.e., retaining the association with the *cursus publicus*, while accepting the implications of the grammar).[77]

Given these weaknesses, a more popular alternative has been one or both parts of the ingenious double interpretation proposed by August Oxé in 1926.[78] Understanding both *miliarios* and *tabelarios* as adjectives agreeing with the understood noun *lapides*, he translated the former as 'milestones' and the latter as 'intermediate' or 'stade-stones', that is, stones marking the stades between the milestones. These, he argued, were small, quadrangular stones, marked with letters to the value of the numbers, reminiscent of voting-tablets (*tabellae*); accordingly, the name *tabellarius* was understood by popular etymology in the sense of 'messenger stone', since they also allowed horse-riders to mount without help. A passage of Plutarch's life of Gaius Gracchus (7.4) was adduced in support (apropos of the beauty and utility of his road-building schemes): 'In addition, the whole road was measured off in miles, and stone columns set up as markers of the distance. Other stones were set up on both sides at lesser intervals to enable riders to mount their horses without the need for further assistance.' The 'mounting-block' theory was soon endorsed by Albert Grenier, while that of the 'stade-stone' has gained widespread support, including that of Ernst Badian in the second edition of the *Oxford Classical Dictionary*.[79] However, no such stade-stone has ever been identified archaeologically.

In the meantime, rejecting Hirschfeld's solution as logically incoherent and objecting both historically and linguistically to the messenger interpretation, Max Cary added a new twist to the debate by proposing that the missing noun to be understood with *tabelarios* was *stipites* (wooden posts).[80] He thus thought of them as 'sign-posts with wooden arms (*tabellae*)', whose perishability neatly explained their archaeological invisibility. As a historical parallel he drew attention to a report of Megasthenes (preserved in Strabo, *Geog.* 15.1.50) to argue that direction-indicators were a feature of roads in the Hindu empire of Chandragupta: 'they built roads and every ten stades erected an upright stone and displayed the turn-offs and distance between'. According to this interpretation, the side roads were not marked at the actual forking but on the nearest milestone. Thus Cary suggested that the Romans went one better and created 'direction-posts' at the actual turning-off points themselves, a solution that gained the influential approval of, among others, the compilers of the *Oxford Latin Dictionary*.[81] René Rebuffat has even gone as far as to combine the theories of Hirschfeld and Cary to postulate wooden stade-markers.[82] However, not the least of the difficulties of Cary's theory is that it requires different nouns (*lapides* and *stipites*) to be understood with the two adjectives *miliarios tabelariosque*, which seem so closely linked.

More recently, Albert Deman concurred in considering the *tabellarii* of Polla as inanimate. But, rather than interpreting *tabelarios* with *lapides* as understood, he preferred to see it as denoting a variety of shelter supposedly erected beside the road in association with the land distribution schemes in order to house administrative archives; edifices where travellers might consult a *tabula* or *tabella*, showing the road connections of the region.[83] However, his argument was vitiated by its reliance on supporting evidence of dubious authenticity and validity.[84]

None of these solutions seems entirely satisfying. More promising is that proposed by Giancarlo Susini in 1984 and taken up with alacrity by Vittorio Bracco in a brief *notizia* in *Epigraphica* the following year.[85] This was founded on a careful reanalysis of the entirety of the Polla text. For one of the principal difficulties the stone presents in assessing its import is its uniqueness in terms of genre. This has to be addressed before any proper assessment of its content can be attempted. As Gian Piero Givigliano recently stated, it is possible to interpret this inscription according to several categories simultaneously.[86] As well as being a simple commemorative monument, labelling the anonymous magistrate's construction of the road, forum and public buildings that were all presumably visible from its original emplacement, it comprehends a description of the distances between that lost location and the principal places along the new road, as well as an advertisement of his extraneous activities. Indeed the slight stepping out and indentation of the text at various points would have encouraged the contemporary reader to break the document down into separate sections.[87] Its variegated nature explains the range of terms by which it has been known, none of them entirely satisfactory.[88] The Polla text (leaving aside the list of distances) is certainly reminiscent of the genre of the inscribed *elogium* that the Romans erected to commemorate the deeds of heroes or ancestors. These served as labels (*tituli*) to accompany the *imagines* (statues or portrait busts) that adorned public spaces and the *atria* of aristocratic households.[89] However, Gianfranco Tibiletti more rightly christened the Polla text an *autoelogium*.[90] For, while *elogia* too name their subject in the nominative (as is the grammatically necessary supplement for the missing name here), the crucial difference is that they are uniformly expressed in the third person. The Polla text is extraordinary in being expressed in the first person. This is unusual for any commemorative text and makes it particularly difficult to construe the stone as a statue base, despite its having been described as such in the earliest reports.[91] In order to defend identification of the stone as a base, one might either take refuge in arguing that its relatively early date invalidates the application of such a rigid typology or argue that this text's grammar has been assimilated to that of the more familiar *miliarium* (milestone), with the erection of which it is, after all, associated.[92] Unfortunately, on the rare occasion that milestones do explicitly carry a verb of authorship, this is invariably in the third, not the first, person and, while the tone of the Polla text is admittedly boasting, it is hard to imagine that the cultural climate of second-century Polla could be so different from the Graeco-Roman mainstream that such self-glorification would seem anything but an act of unparalleled hubris.[93] Nevertheless, such considerations did not prevent Vittorio Bracco from imagining this stone as but one of a series supporting the magistrate's likeness. He even endorsed Ernst Lommatzsch's implausible theory that the statue at the straits was the road-builder's own, though the bald labelling *ad statuam* ('at the statue') surely indicates that a pre-existing and already well-known landmark is intended (the statue of a god or hero seems most likely).[94]

Giancarlo Susini has cogently argued that the Polla text should be seen as a composite. His chief distinction was between the elements of *elogium* (*uiam fecei . . . poseiuei, et eidem . . . heic fecei*) and the table of mileages sandwiched between them (*hince sunt . . . meilia CCCXXI*), a distinction reinforced by the observation of the variation in linguistic usage: *ab Regio* (line 1) but *af Capua* (line 9).[95] As Hermann Dessau long ago pointed out, in many ways the closest parallel for the self-laudatory element is Augustus' record of *res gestae et impensae*, that very personal document of deeds and spending that the emperor ordered to be inscribed after his death on bronze pillars outside the entrance to his mausoleum.[96] Both texts are similarly extraordinary in addressing the reader in the first person and, I think, likewise both documents should be viewed as attempts, if not to fix, certainly to influence strongly the way in which their authors would be remembered by posterity. In his list of mileages and precise figure for returned runaways, the author of the Polla text certainly shares the same delight in statistics as his more famous counterpart. Augustus' *Res Gestae*, of course, includes a claim to road-building activity, which significantly contrasts with the passive-voiced wording of the public monument commemorating that very same activity.[97] Thus it seems justifiable to adopt Susini's suggestion to term the Polla text *Res Gestae* or, following Suetonius' categorisation of Augustus' monument, an *index* (catalogue) *rerum a se gestarum*.[98] Susini considered the *elogium* elements of the Polla text to have been straightforwardly copied from a pre-existing document. I prefer to understand the wording as the magistrate's wishful *Vorarbeit*, or rehearsal, for his posthumous *elogium*, representing perhaps the highlights of a boastful speech delivered at the hand-over ceremony for his building works in Polla. Indeed, it is not impossible that this is a posthumous memorial. This would render the first-person narrative less anomalous. However, the seductive parallels with Augustus's document need not lead us to suppose a similarly widespread distribution for the Polla text. It is surely the anonymous magistrate's particular connection with this town, outlined in the last line, that has prompted the commemoration of his deeds here.

Susini's distinction of the sections of *elogium* and intruding table of distances (*tabula miliaria*) led him and Bracco to the realisation that this latter section, if taken on its own, might legitimately be termed *tabellarius*, deriving it from *tabella* as the diminutive of *tabula* in the sense of a board for public notices, a register or list.[99] This solution is very attractive. For, if this is right, the insertion of the itinerary list itself might be considered almost as a parenthesis explicating the content of the '*tabelarios*'. Furthermore, it is easy to imagine the places named in this list as the likely locations of the other *tabellarii*. Clearly it would seem that, as with the *miliarii* (*lapides*), their designation has arisen from their salient feature. In the case of milestones it was their indication of the mileage, in the case of the *tabellaria* – as it seems legitimate to coin them on the analogy of the development of *miliarium* as a free-standing adjectival noun – it is their listing, tabulation, of the stages along the road with an indication of the distance. Moreover, the existence of the toponym 'Tabellaria' (a stage on the via

Aurelia between the crossing of the River Minio and that of the Marta in southern Etruria) may provide an ancient attestation, if we assume that a set of itinerary stones was the most notable feature of the location; a phenomenon equivalent to that of the many places which took their designation from their milestone, for example, *ad sextum*, *ad vicesimum*, and so on.[100]

TABELLARIA AND *ITINERARIA*

The failure of this neat solution to the *tabellarius* conundrum to supplant the previous interpretations lies, I believe, in a failure to seek out corroborating evidence or explore the full consequences of this identification for the history of Roman travel and our perception of the availability of geographical knowledge to the Roman public. From the context of their mention on the Polla stone it is clear that *tabellaria* must have constituted a standard and recognised feature of the infrastructure of Roman roads alongside bridges and milestones. Their close linking with the latter naturally reflects the interdependent nature of their roles as aids to the traveller, the one in planning a journey, the other in assessing its progress. Both were provided by a road's builder, the data for the *tabellaria* deriving from the survey conducted by the engineers in laying out the road. Thus the details of, at least, the major stages along a Roman public road was a matter of public record from its inception and remained so. As the very date of the Polla example demonstrates, there no longer remains any need to postulate a dependence on the archives of the *cursus publicus* for the raw material of the Antonine Itinerary or *Tabula Peutingeriana*. However, if *tabellaria* were such standard features of Roman roads why have they not been noticed before? After all, over 4,000 examples of milestones survive from right across the empire. In fact, on the basis of our definition of the *tabellaria* alluded to by the Polla text, at least eight surviving examples of these artefacts can be distinguished in the archaeological record, from places as far apart as Spain and Lycia.[101] Despite considerable variations in physical form – some are plaques, others circular columns, one an eight-sided pillar – the common factor is the *tabella* of route-stations clearly displayed to public view. This may seem a paltry total but, unlike milestones, they need not always have been in such a durable medium as stone, since the foundation of new towns or re-engineering of the courses of roads may have necessitated periodic revision. In addition certain factors contingent on their likely usual positioning may have rendered them particularly liable to destruction.

Certainly until recently the generally recognised extant examples of these itinerary stones have been fragmentary, or hypothesised on the basis of votive models; moreover, their distribution seemed to be confined to the western part of the empire: four from the Gallic provinces, one from Africa and two from Spain. All may be identified as *tabellaria* from the shared characteristic of listing places, some along a single route, some along a series of routes in the region, and others distances radiating outwards from their location. The most impressive monument was that which appeared to be attested as having existed at Gades, on the basis of four columnar silver beakers, possibly Augustan in date, that had been thrown into the sacred spring at Vicarello near the Lago di Bracciano in southern Etruria (*CIL* XI 3281–4).[102] They vary in size (from 9.5 to 15.3 cm) and workmanship but all are similar in the respect that they detail the stages from Gades via Gallia Narbonensis to Rome in lists disposed into four panels divided by mini-Corinthian pilasters, as well as advertising the total distance.[103] Moreover, on one the stages are broken into six journeys: from Gades to Hispalis, from Hispalis to Corduba, from Corduba to Tarraco, from Tarraco to Narbo, and from Narbo to Augusta Taurinorum, before concluding the itinerary to Rome:

A Gadibus Roma<m>	
ad Portu	*XXIIII*
Hasta	*XVI*
Vgiae	*XXVII*
Orippo XXIIII	
Hispali VIIII	
Ab Hispali Cordubae	
Carmone	*XXII*

It is easy to imagine these as modelled on a real monument in Gades, which might have been similarly circular, or rectangular, or simply a series of plaques on a wall.

As for surviving monuments, part of the route outlined on the Vicarello beakers, that along the east coast from Valentia (Valencia) to Tarraco and beyond, was inscribed on a stone found at Valentia; though, as it survives, it lacks any distances (*CIL* II 6239= II^2 14, 38). More elaborate itinerary inscriptions are known from north of the Pyrenees. At Augustodunum (Autun) there survived three broken slabs from what seems to have been an imposing monument *c*. 3 m high by 2 m wide (*CIL* XIII 2681a, b [= *ILS* 5838], c = XVII/2 490a–c).[104] The first, now lost, listed stages between Bononia and Fines Gallorum on the via Aemilia in northern Italy, the second lists two routes from Autessiodurum (Auxerre), and the third traces a route northwards from the Saône to the Moselle since it mentions Andemantunnum Lingonum (Langres) and Tullum Leucorum (Toul). Given that for the Gallic sections the distances are still measured in miles, this monument ought to date to before *c*. AD 200. Of slightly later date is the fragment of three faces of an originally octagonal column from Atuatuca Tungrorum (Tongeren, near Maastricht: *CIL* XIII 9158 = XVII/2 675 = *ILS* 5839), listing four separate itineraries, all measured in leagues. As it survives, the first face traces a route southwards up the Rhine from Rigomagus (Remagen) to Borbetomagus (Worms), the second a route from Durocortorum (Reims) north-eastwards to Samarobriva (Amiens), and the third two routes from different locations in Picardie, Castellum Menapiorum (Cassel) and Bagacum (Bavay):

ITEM
A CAS-
TELLO
Fines Atrebatium l. XIIII
Nemetac(um) l. {---}
ITEM
A BA{G-
ACVM}

Also from the Low Countries, a broken stone from Buchloche near Junglinster in Luxembourg (*CIL* XIII 4085 = XVII/2 676) lists on three faces a number of routes emanating from Moguntiacum (Mainz) on the Rhine, nearly 150 km to the east.

A particular subset of *tabellaria* comprises two columns, one from Gaul and one from Africa, that, instead of giving stages along routes from A to B, list a set of figures for distances from their findspot to other locations at various points of the compass. Thus these stones mark crossroads (*compita*); that from Allichamps near Bruère, Cher (*CIL* XIII 8922 = XVII/2 489 = *ILS* 5837), listing the distances to Avaricum (Bourges) to the north, Mediolanum (Chateau-Meillant) to the west, and Neriomagus (Néris-les-bains) to the south-east; and that from Fedj-Souïoud, but now in Aïnbeida, Tunisia (*CIL* VIII 10118 = 22247 = *ILS* 5836), dating from the reign of Elagabalus (AD 218–22), which gives distances, also in counter-clockwise order, to Carthage in the north-east, Hippo Regius to the north, Cirta to the north-west, Lambaesis to the south-west and Theveste to the south-east.

All these examples have long been known but our corpus of *tabellaria* has recently received a most impressive addition that is in several respects reminiscent of the *elogium* of Polla: the so-called *stadiasmus prouinciae Lyciae* of AD 45 from Patara on the south coast of modern Turkey (*SEG* 44 [1994], 1205). By reassembling the fifty-three separate inscribed blocks found built into the town's later Roman walls and scattered in marshy ground before them, a rectangular stone monument can be reconstructed, which measured 1.6 m wide by 2.35 m deep by 5.5 m tall, inscribed in Greek on three of its four sides. On the narrow front face it bears a dedication to the emperor Claudius thanking him for exercising his providence in liberating the Lycians from disorder, lawlessness and piracy (i.e. a euphemism for annexation to the empire), which suggests that it may have formed the basis of an equestrian bronze statue of the emperor. On the left-side face begins a separate text recording Claudius' order, through his legate Q. Veranius, Lycia's first governor, to survey the province's roads, which then duly goes on to describe the province's road network (radiating out from Patara) in the form of three consecutive itineraries, all taking

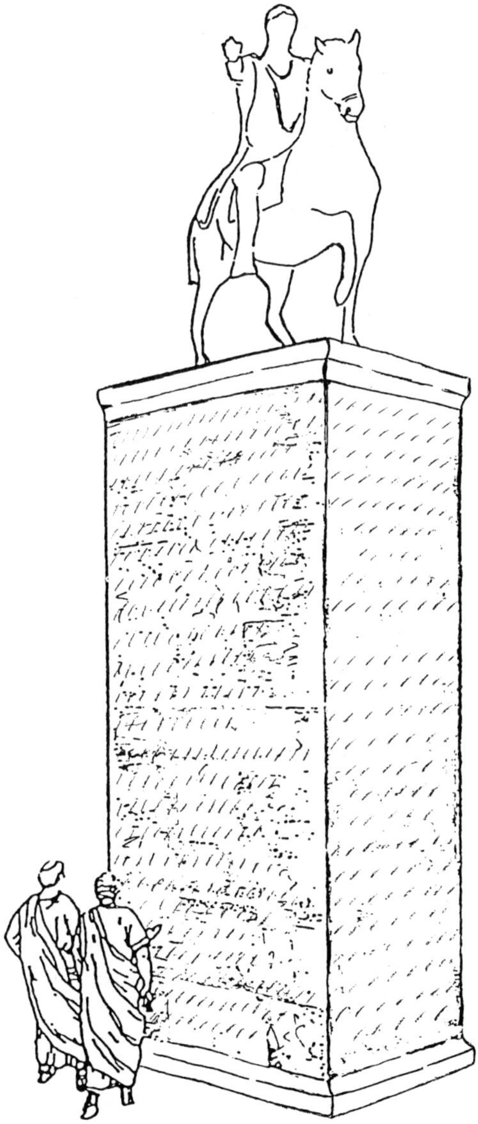

Figure 3.5 So-called *stadiasmus prouinciae Lyciae* (sketch by R.W.B. Salway after Şahin)

Patara as their starting point, but ending respectively at Kaunos along the coast to the west, Kibyra inland to the north (both in Asia), and Attaleia along the coast to the east (in neighbouring Pamphylia); a list which extends on to the right-hand side. Even from the short excerpts in the preliminary publication it is clear that these itineraries are expressed in just the same way as the Latin texts, with the difference that the distances are given in stades. This detail in fact suggests that the figures recorded here relied on existing local information rather than being the result of the survey signalled on the monument, since Roman engineers would undoubtedly have used the mile as their measurement of distance. And in fact the road-building activity was still very much under way under Veranius' successor T. Clodius Eprius Marcellus *c.* AD 50–4.[105]

The almost complete annihilation of even this huge *tabellarium* is instructive. It is plausible that, where *tabellaria* existed, they tended to be located at the gates of towns. Certainly this was the case in Valentia and Atuatuca Tungrorum, where in the latter instance the surviving slab was also recovered from the fabric of the town's later Roman walls.[106] This may in part explain the apparently low survival rate of this category of monument in comparison with the *miliaria*. Thus it seems less improbable that *tabellaria* were a reasonably widespread phenomenon, from which convenient handlists of stages might easily be compiled; these handlists would in turn, I suggest, be the ancestors of our manuscript itinerary collections.

CONCLUSIONS

There is no need to invoke governmental files as the immediate source of such documents as the *Itinerarium Antonini* and *Tabula Peutingeriana*. Instead, their variegated nature reflects the fact that they were based on the compilation of itinerary data from publicly displayed lists of stages and distances (*tabellaria*), erected at disparate locations at different times and no doubt slightly variant in the level and type of detail they offered. As such the amassing of this data is very much rooted in the experience of travel rather than the theory of geography. Once in manuscript such lists could well go on being copied long after the monumental texts from which they were derived had perished or been replaced by more up-to-date versions. Thus, while assembled in its final form with an eye to geography, the internal structure of the collection known as the Antonine Itinerary is explained by its gradual accretion from disparate similar collections, its coverage representing the random factor of the areas with which its various compilers had connection.

Both the Antonine Itinerary and Peutinger Table can, then, be appreciated as multi-layered documents composed of a mixture of itineraries actually travelled as well as those simply recorded for future reference. The process can be seen at work in the Rome–Hiera Sycaminos section of the Antonine Itinerary, of which, given its annotations, the stretch from Rome as far as Singidunum must derive from a specific journey made (124, 8–132, 1). As was noted long ago, this incorporates at Ariminum a sort of parenthesis giving an alternative route to Aquileia than that taken, indicated as such by the lack of glosses (126, 5–9).[107] This surely derives from the passing traveller having noted down the alternative route from a *tabellarium*, perhaps at Ariminum. The significance of the clustering of routes is also thus brought into focus. These surely represent the location of a *tabellarium* from which a traveller has taken down not simply the route he intended to travel but all the possible routes listed from that place. In fact, a single *tabellarium* may frequently have been even more informative than that. For, although apropos of the Autun stone Ernest Desjardins described the itinerary monuments as *poteaux-étoiles de routes*, even in their fragmentary state it is clear that several of our surviving *tabellaria* (Autun, Tongeren, Junglinster) exhibited a more complex pattern, since their find-spot does not necessarily feature as the *caput viae* of the majority, or even any, of the routes attested.[108] This suggests that more elaborate examples such as these might contain details of the route network extending over a whole region as well as for the even longer trip from that region to Rome. By collecting just a few such *tabellaria* in his collection, a traveller could swiftly build up an extensive knowledge of the routes pertaining to quite distant areas.

This still begs the question as to the nature of the potential traveller who might make use of a *tabellarium*. The need met by *tabellaria* is not obviously that of the regular traveller for reasons of commerce, or business on behalf of the state, or in connection with the economies of the landed estates of the wealthy. In those cases, the information necessary would no doubt have been passed on by experience or through official channels. Rather, these instruments of public display meet the needs of the private individual forced occasionally to make journeys beyond the region with which he or she is normally familiar. Such a constituency will have changed over time. For instance, at the time of its erection, the author of the Polla stone may well have envisaged the Roman citizens now sprinkled across southern Italy as a group whose access to Rome in order to vote might be facilitated to his distinct advantage. Obviously the advent of the principate will have put an end to this motive but the tide of litigants seeking out Roman magistrates will have in no way been stemmed. In this regard the publicly displayed lists of distances will have served a crucial function in fixing – in combination with the standard calculation of twenty miles per day – the agreed time for appearance in Rome or, in the provinces, at the current assize of the governor. The imperial court generated yet more business in a similar vein, as the destination of increasing volumes of public embassies and private appellants.[109] The pretext for Theophanes' journey was certainly business at the regional capital of Antioch and the original impetus for the travels of the Anonymus Burdigalensis may well also have been similar, this time with the imperial court as its goal. Might not the unfortunate who came to rest at the Vigna Codini columbarium have been an easterner who died in the pursuit of a similar mission to the imperial court of their day?

In this light the chronology of the development of our manuscript itinerary collections might also be significant; it is to be remembered that the source materials of the Antonine Itinerary and the *Tabula Peutingeriana* seem to have reached their final forms by about AD 300 and 350 respectively. Obviously in either roll or later codex form such collections could convey much more information than could practically be contained on any single *tabellarium* and as such provided material for the planning of a far greater variety of journeys beyond one's own region. As indicated by the Cadiz and Autun examples, a provincial *tabellarium* might regularly provide an itinerary for travel to Rome. However, by the later third century, the imperial court or, rather, courts were rarely if ever in Rome. Instead they had become peripatetic entourages, whose residence over the medium to long term was not always predictable. In this world, those on personal or civic missions needed to plan and undertake more complex journeys in pursuit of the court. Given its coverage, although largely useless for the leisured pursuit of cultural tourism in the Aegean and Asia Minor, the Antonine Itinerary would have served this more businesslike purpose. Thus, I suggest, the needs of such business were the spur to the building-up of collections such as the Antonine Itinerary and that which underlies the *Tabula Peutingeriana*. And, indeed, even if it was only ever a display model, the *tabula* itself would have been a useful tool in negotiating the complexities of this particular late antique game.

NOTES

1 See most recently the remarks of Arnaud (1988) 303; (1993) 33; and Laurence (1999) 85.

2 Cuntz (1929), reprinted with updated bibliography by G. Wirth (1990). Both are conventionally cited by the page and line numbering of Wessely (1735). For the Peutinger Table (Vienna, Nationalbibliothek MS. Lat. 324) the definitive edition is Weber (1976), but it is most conveniently consulted in, and cited according to, Miller (1887) (repr. Brockhaus, Stuttgart, 1962).

3 Capelli and Pesando (1991) 41–4; to which list now add Şahin (1994) 130–7 = *SEG* 44 (1994), 1205.

4 *CIL* VI 5076. For Theophanes, see *P.Ryl.* IV 627–8, 638. The only edition of the *Itinerarium Burdigalense* to take account of all four known manuscripts is that by Glorie (1965) 1–26.

5 The *bematistai* (pacers) who recorded the itinerary of Alexander's conquests (Pliny, *NH* 6.61) have left no trace in the Greek tradition. Similarly, among the seventy-seven books written by Hermogenes of Smyrna were a *stadiasmus Asiae* and a *stadiasmus Europae* (*CIG* 3311 = *IGRR* IV 1445) but their nature and content is unknown.

6 Müller (1855) 244–54.

7 On the 'directional' locative see Adams (1977) 38; Mackay (1999) 229–39. *Contra* Pallu de Lessert (1904–5) 115–38 and Heurgon (1952b) 169–78, who,

ignoring the syntax, consider the names in itineraries to have already become fossilised indeclinable nouns.

8 Migne (1879) col. 1127; Kubitschek (1919) 4–5.

9 As remarked already by Tozer (1897) 306–8.

10 *Latina* [sc. *uia*] *ab urbe . . . Compitum m.p. XV, intrat in Lauicanam* (305, 7–306, 2); '*Beneuento* [sic], *mansionibus quibus et in Prenestina, m.p. CI*' (305, 5–6).

11 Having accidentally omitted the crucial '*uiam*', the scribe of the *codex Escurialensis* (R II 18) had to go back and squeeze it in before the main text in smaller letters; on the practice of this scribe see Cuntz (1893) 260–98.

12 Arnaud (1993) 33, fig. 1; Brodersen (1995) 186.

13 Schnetz (1940) repr. with index added by M. Zumschlinge (Stuttgart, 1990) 1–110. Miller (1888); (1916) xxvii–xxix; Schnetz (1942).

14 Mommsen (1851) 96–103; Wartena (1925) 46–56. Cf. Schillinger-Häfele (1963) 238–51; Dillemann (1997) 27.

15 Kubitschek (1902) 81; (1914) col. 2336; (1919) cols 2101–18; followed by: Weber (1976) 22–3; Bosio (1983) 158–9; André and Baslez (1993) 399.

16 Brodersen (1995) 268–87. The prime advocate of 'hodological space' is Janni (1984). Cf. now Engels (1999).

17 Dilke (1985) 31; Sechi (1990) 206.

18 Pliny, *NH* 3.16–17, 6.139; Reise (1878) repr. Olms, Hildesheim, 1964; Anon., *Divisio orbis terrarum*, p. 20 = Baehrens (1883) 84; Eumenius, *Oratio pro instaurandis scholis* 20.2 = IX (4) in Mynors (1964) 242.

19 Detlefsen (1909) repr. Rome, 1972; Klotz (1931) 38–58, 386–466; Dilke (1985) 50, apropos of Pliny, *NH* 6.139.

20 *Pace* Nicolet (1991) 102–3.

21 Gallazzi and Kramer (1998).

22 Ibid. 199. See also Brodersen (1999), ii–iv.

23 Which should serve as warning that, whatever the correct form for the name of the ancient site excavated at Torre Annunziata, it is not likely to be 'Oplontis'; cf. Dilke (1985) 114, with n.11.

24 Laurence (1999) 86–7, with fig. 6.4.

25 For example, Stolte (1949) 26; Bekker-Nielsen (1988) 157.

26 Schweder (1893) 485–512 argued that it was purely a display artefact.

27 For alternative, and quite differently nuanced, translations cf. Dilke (1987) 236–7; and Milner (1993) 71.

28 Lieb (1974) 31–4; Weber (1976) 9.

29 For example, Anon.Rav. IV 25 [60, 10–12]: Maguntia (modern Mainz), Anternacha (Andernach); IV 26 [61, 27–46]: Sphira (Speyer), Stratisburgum (Strasbourg), Bazela (Basel), Wrzacha (Rorschach), Ziaberna (Rheinzabern), Ziurichi (Zürich).

30 Gaius, *ad edictum provinciale* 1 (*Dig.* 2.11.1), Ulpian, *ad edictum* 2 (*Dig.* 50.16.3 pr.); and elsewhere Ulpian (*Dig.* 38.15.2.3) and Modestinus (*Dig.* 27.1.13.2) equate twenty miles with one day's travel.

31 As emphasised by Laurence (1999) 84–94. Rutilius Claudius Namatianus, *De reditu suo siue Iter Gallicum* (ed. Doblhofer (1972–77)) II.7–8: *Interualla uiae fessis praestare uidetur / Qui notat inscriptus milia crebra lapis.*

32 Susini (1992) 119–121.

33 Kubitschek (1919a) 20–6; Hunt (1982) 58–9.

34 Theophanes' entourage comprised at least seven, and aristocratic pilgrims seem to have been accompanied by at least as many: Rees (1968) 171; Hunt (1982) 74–81.

35 Petschenig (1913) 83; *Sermo* 5.2: *Eliguntur ciuitates, in quibus triduum, quatriduum, et plures interponantur dies, si aquis abundant, commerciis frequentantur*; on which see further below. *P.Ryl.* IV 627 verso is covered with the memoranda of Theophanes' accounts; on his likely use of the *cursus publicus* see Roberts's commentary in *P.Ryl.* IV, pp. 104–7.

36 Casson (1974) 192.

37 *Pace* Bekker-Nielsen (1989) 148–61.

38 Baehrens (1886) 150–5, Lucilius' frags. 71–117; André and Baslez (1993) 95–7. See generally Skeel (1901) 81–109; Chevallier (1972) 228–32.

39 In *P.Ryl.* IV 638 Botrys appears before Biblos, whereas on the way out Theophanes travelled directly from Biblos to Tripolis on day 20 (*P.Ryl.* IV 627 col. vii 325).

40 In two of the four manuscripts – Madrid, Archivo Histórico Nacional 1279 (1007 D [254]) and St Gallen 732 – the itinerary has been deliberately discarded to preserve this section alone.

41 On the nature of the comments see Douglass (1996) 313–33; with Weingarten (1999) 291–7.

42 That is, *inde ascendis Matronam* (556, 1), *transis pontem, intras Pannoniam inferiorem* (561, 5), *intras Pannoniam superiorem* (562, 8), *a Constantinopoli transis Pontum, uenis Calcedoniam, ambulas prouinciam Bithyniam* (571, 9–10), *ut ascendas Sion* (591, 7), *et uenis Odronto* (609, 5).

43 Doubted for the Bordeaux traveller by Hunt (1982) 57–8.

44 See *P.Ryl.* IV 623 with Moscadi (1970) 88–154. On the nature of journey see Rees (1968) 180.

45 Cf. the attempt by Douglass (1996) 327 n.61 to argue a different interpretation of the dates in order to locate the pilgrim in Jerusalem for Easter.

46 Barnes (1982) 79.

47 Berchem (1937) 117–202; (1973) 123–26; and (1974) 301–7; Levi and Levi (1967) 97–102; Chevallier (1972) 27; Reed (1978) 228–54; Janni (1984) 31; Cosme (1998) 93.

48 Dilke (1985) 115.

49 Thomson (1948) 376; Sherk (1974) 560; receipt of *itinerarium* as a guide to travel and location of victualling points: Janni (1984) 31; *Soldaten erheilten für ihre Marschroute einen* Texte *mit einer Liste der auf ihrer* route *jeweils zu erreichenden* landmarks: Brodersen (1995) 173.

50 Brodersen (1995) 174: *weil der Kaiser nur für die alles vorzubereiten befiehlt, die (dem* itinerarium*) folgen und nicht links oder rechts vom vorgeschriebenen Weg abweichen.*

51 Kubitschek (1891) 179, stemma 209.

52 Provincial headings: <iter> prouinciae Africae (2, 1), iter Sardiniae (78, 5), Corsicae (85, 4), Siciliae (86, 2), Italiae (98, 2), iter Thraciae (175, 1), iter Britanniarum (463, 3).

53 Biblioteca del Escorial MS. R II 18, in a late seventh-century Visigothic uncial hand; Lowe (1966) 1631.

54 Arnaud (1992) 374–79.

55 Levi (1971) 4; Habicht (1985) 19–20.

56 For example, Item de Pannoniis in Gallias (231, 8), De Dalmati in Macedoniam (337, 3), De Italia in Gallias (339, 6).

57 For example, Samarobrivae (379, 10; 380, 1), cf. Ambiani (362, 4). The latest surviving milestone from the Gallic provinces is *CIL* XIII 9150 = XVII/2 575 from Antunnacum (modern Andernach) of AD 164, while the earliest 'league-stone' is *CIL* XIII 9137 = XVII/2 558 from Tolbiacum (Zülpich) of AD 202.

58 Paris, Bibliothèque Nationale MS. Gr. Suppl. 1354, no. 5.; Dilke (1985) 121; Rebuffat (1986) 85–105; Arnaud (1989) 373–89.

59 For example, Dionysius 'Periegetes', *Orbis Descriptio*, which spawned later Latin versions by both Rufus Festus Avienus and Priscianus: Müller (1855–61) II, 104–76, 177–89 and 190–9.

60 On the favoured destinations of the latter see Casson (1974) 138–48.

61 Cf. Miller (1916) xxx; Leylck (1993) 203–6; French (1996) 101.

62 Arnaud (1988) 309.

63 On the icons see Levi and Levi (1967) *passim*; Dilke (1985) 193–5.

64 Bühl (1995) 131–2. Ausonius, *Ordo urbium nobilium*: Roma, Constantinopolis, Carthago, Antiochia, Alexandria, Treueri, Mediolanum, Capua, Aquileia, Arelate, Hispalis, Athenae, Catina, Syracusae, Tolosa, Narbo – and last but not least – Burdigala.

65 Cf. Weber (1989) 113–17, who dates the archetype to the early fifth century. Levi and Levi (1967) 147 prefer to put the redactor in the east because of prominence of Antioch and the Orontes.

66 Cf. the more ingenious explanations of Chevallier (1972) 39–40; Mann (1987) 285–6; Black (1995) 98–106.

67 Casson (1974) 172, cf. 186; Chevallier (1972) 46–54; Dilke (1985) 122–4; André and Baslez (1993) 399; Brodersen (1995) 179–84. The wall souvenirs are *CIL* VII 1291 = *RIB* II/2 2415.53 and *Ann.Epigr.* 1950, 56, on which see Heurgon (1951) 22–4.

68 Itinerary stone: Kubitschek (1914) cols. 2314–19; Casson (1974) 172; cf. Mari (1996) 250–1.

69 *CIL* I 551=*CIL* I^2 638=*CIL* X 6950=*ILS* 23=*ILLRP* 454=*Inscr. It.* III 1, 272; photograph in Degrassi (1965) 135, no. 192a. On the route see *It.Ant.* 109, 1–111, 5 and (the somewhat confused) *Tab.Peut.* VI 5–VII 1; cf. Ruoff-Väänänen (1975) 93–8.

70 Susini (1984) 110; Bracco (1985) 97.

71 Accordingly the bibliography is as enormous as it is diverse, the text being cited already by Freccia (1554) 234, apropos of the history of pastoralism in Italy (I am indebted to Prof. M.H. Crawford for this); for a reasonably recent round-up of the bibliography see Solin (1981) 55–7; Degrassi and Krummrey in *CIL* I^2 638, fasc. iv (1986) 922–3; Crogiez (1990) 391–3 nn. 6–11; and Givigliano (1994) 347–8 nn. 200–14. For publications prior to the 1880s see the commentary of Mommsen in *CIL* X (1883) p. 707.

72 This is the usage attested frequently in Cicero's works (e.g *Epp. ad fam.* 9.15.1, 10.31.4, 15.18.2, *Phil.* 2.31.77, *de prov. cons.* 7.15) as well as by Livy (45.1.6), clearly having been formed with the *-arius* adjectival suffix from *tabellae* in the sense of writing-tablets.

73 Mommsen (1856) 141–8, cf. *CIL* I (1863) no. 551, pp. 154–5. For examples of rectangular milestones see, e.g., *CIL* I^2 457, 459, 647, 651 and 2877.
74 Hirschfeld (1907) 169–70 = (1913) no. 44, 708–9.
75 Sandys (1919) 132–3; Holmberg (1933) 36–7; Wiseman (1970) 151; Brodersen (1995) 172 with n. 4.
76 Desjardins (1878) 52–3; Humbert (1887) 1648 with n. 65; Ramsay (1920) 79–86; Schroff (1932) col. 1845; Reincke (1935) col. 1525; Levi and Levi (1967) 102; Radke (1973) col. 1446; Gross (1975) cols 475–6; Crogiez (1990) 389–431.
77 Stockton (1981) 18.
78 Oxé (1926) 219–20, 242.
79 Grenier (1931) II, 79–81; Schneider (1935) col. 400; Pflaum (1940) 54–7, 242–5; Degrassi (1957) no. 454, p. 254 n. 3; Calabi Limentani (1968) 290; Pekáry (1968) 65–7; Badian (1970) 1032; Chevallier (1972) 37–8.
80 Cary (1936) 166–7. He found Hirschfeld's theory grammatically unimpeachable but unconvincing, seeing that the Polla inscription itself (at no less than sixty-eight words, being one of the longest of its class) is carved into stone.
81 Warmington (1940) 150–1; De Rosalia (1978) 122; Glare (1982) 1879, s.v. 'tabellarius', whence Yavetz (1988) 43, and Lomas (1996) no. 113, 73–4.
82 Rebuffat (1995) 134.
83 Deman (1976) 32–7.
84 Relying for support on the wording of *CIL* XII 5732 from Antipolis (Antibes): *Adi viator si libet intus veni / tabula est aenea quae te cuncta perdocet*, which, indeed, is likely to be a fake of the time of François I; on which see Sherk (1974) 560 n. 87. Deman's citation of the mention of *viae*, *tabularii* and *pontes*, which appear in close proximity to each other in a law of Jean I (1316) or Jean II le Bon (1351) on road infrastructure, is a 'red herring' because these are *tabularii* (not *tabellarii*), even if he could determine what they were, which he cannot.
85 Susini (1984) 103–10, and Bracco (1985) 93–7.
86 Specifically those of *miliarium*, *itinerarium* and *elogium*: Givigliano (1994) 289.
87 Susini (1984) 103–10.
88 In addition to *elogium*, it has also been called variously the *lapis Pollae*, *tabula Popilliana* and most recently 'Il *tabellarius* di Polla' by Bracco (1985) 93–7.
89 A rare example of the text of a public *elogium* to survive from the republican period (albeit in a later copy) is *CIL* I^2 25 = *ILLRP* 319, the marble statue base for C. Duilius cos. 260 BC from the Forum Romanum. Nevertheless, the cycle of *elogia* from the Forum Augusti reflects the genre; on which see *CIL* I^2, pp. 187–201, and the commentary of Degrassi (1937). On the distillation of epitaphic *elogia* and *tituli imaginum* from funerary *laudationes* see Flower (1996) 179–84 and 274–5.
90 Tibiletti (1955) 255, followed by Panebianco (1963–4) 3–22 and Susini (1984) 103–10.
91 The text is prefaced with the label '*in principatu ad basim*' (i.e., in the Principality of Salerno, (?) on a base) in its first known recording (some time between 1477 and 1484) in the contemporary and mutually dependent epigraphic manuscripts of the north Italian humanists Michael Fabricius Ferrarinus and Iohannes Iucundus, from whom Konrad Peutinger took over the description wholesale in the epigraphic manuscript used as the source of its first

printed publication by Petrus Apianus and Bartholomaeus Amantius (1534) 104.

92 That found at Vaccarizzu near the village of Sant'Onofrio (near Vibo) on this very road is typical of the genre in commemorating its author in the nominative while leaving any verb understood: *CCLX* | *T. Annius T. f.* | *pr(aetor)*; on which see Pitimada (1953) 343 = Degrassi (1965) 454a.

93 A probably almost contemporary example from Asia Minor (Degrassi (1965) 456) provides a verb but is unfortunately inconclusive because lacunose at the crucial point. But imperial examples, for example, *CIL* XI 6668=*ILS* 9497: [Traianus] *uiam Cassiam collabsam a Clusinorum finibus Florentiam perduxit*, demonstrate that the third person is standard. In terms of the commemoration of road-building the closest extant parallel to the Polla text is the series of inscriptions erected by P. Cornelius Dolabella in AD 16/17 and 19/20, recording the opening of new and relaying of existing roads out of Salonae in Dalmatia (*CIL* III 3198–201). Like the milestones, these are in the third person.

94 Lommatzsch (1918) no. 638, p. 510; Luzzatto (1962) 375–98, and Degrassi (1986) 922–3. On the column see Strabo, *Geog.* 6.1.5 with Lasserre's commentary (1967) 221–2; Mercuri (1998) 551–69.

95 Susini (1984) 103–10 considers *af* a dialectical variant; Vine (1993) 188–9, as part of the technical jargon of road engineers.

96 Dessau (1929) 271 with n. 5, where a significant parallel is cited: *CIL* XIV 4012 (Ficulea, Latium), a *cippus* (and *not* a statue base) on which a local benefactor celebrates his road improvements in the first person. On the genre and style of the *Res Gestae Divi Augusti* see Brunt and Moore (1967) 2–3, 6–7.

97 *RG* 20: *consul septimum viam Flaminiam a*[*b urbe*] *Ari*[*minum refeci . . .*]. Cf. *CIL* XI 365=*ILS* 84 (Ariminum): *Senatus populusq*[*ue Romanus* | *Imp. Caesari divi f. Augusto imp. sept.*]|, *cos. sept. designat. octavom, v*[*ia Flamin*]*ia* [*et reliquei*]*s* | *celeberrimis Italiae vieis consilio* [*et sumptib*]*us* [*eius mu*]*niteis.*

98 Suetonius Tranquillus, *de vita Caesarum* 2. *Augustus* 101. 4.

99 Glare (1982) 1898, *tabula* 4–5, and 1897, *tabella* 3.

100 *Tab.Peut.* V 1 (Anon.Rav. *Cosm.* 70.5, 85.52; Guido, *Geog.* 121.28, 130.72); quite possibly to be located at the crossroads with the road from inland Tarquinii to its port at Gravisca, which the *Tabula* puts at '·co·' (i.e. 1 m.p.) from Tabellaria. Cf. Lopes Pegna (1952–3) 386.

101 Excluding *Ann.Epigr.* 1921, 6–9, the tiles from Asturica Augusta (Astorga) in Spain as of dubious authenticity.

102 Heurgon (1952a) 39–50.

103 Drawings in Chevallier (1972) 47, Dilke (1985) 123, Brodersen (1995) 178.

104 On the archaeology of the monument see Grenier (1934) II, 113–18; Thévenot (1969) 59–66.

105 Milner (1998) 120.

106 Respectively found in 1727 at the porta de la Xerca at Valencia and in 1817 outside the porte de Saint-Trond of Tongeren.

107 Elter (1908) 14–16.

108 Desjardins (1893) IV, 21–5. The alternative, in the Autun case, that the stone was removed from an original location in Auxerre, 132 km over the Morvan hills and down the Yonne valley, is not really plausible.

109 Regular appeals from the *prouinciae transalpinae* and *transmarinae* are envisaged by the late second- or early third-century imperial edict recorded on an Egyptian papyrus: *BGU* II 628 recto = *FIRA* I no. 91, 452–4.

4

THE CREATION OF GEOGRAPHY

An interpretation of Roman Britain[1]

Ray Laurence

Recent work on the Mediterranean in antiquity has shown that the land masses bordering the Mediterranean Sea had a particular geography that placed an emphasis on the interaction between its micro-regions.[2] This thesis has suggested that the Mediterranean, unlike other regions of the Roman empire, had a particular type of geography that had its basis in the ecology and interaction promoted by the sea.[3] This raises the question of how Rome viewed its northern provinces and how the north of the Roman empire was interpreted or geographically reorganised to create a version of this Mediterranean geography. Hence, in this chapter, I will be concerned with the view of Britain found in the literary sources after the conquest in AD 43 and in particular the geographical sources of Ptolemy and the Antonine Itineraries. The chapter follows on from work on Roman Italy that set out a thesis that the promotion of mobility, via the building of long-distance roads, created the geographical unity of Roman Italy or *tota Italia* during the late first millennium BC.[4] Here I wish to deal with the creation of provincial geography in Britain to assess how the Roman geographical model was adjusted and altered once outside the context of the connectivity of the Mediterranean and its associated ecology of micro-climates. This might be an appropriate way in which to view the cultural changes often associated with what is known as the *Romanisation of Britain*.[5] Rather than a mass adoption or rejection of the material culture of Rome,[6] we may be seeing a change in the nature of mobility – defined by the circulation of people, goods and capital. It will be argued that the evidence for mobility found in the geographical and historical narratives from antiquity may provide a model for the understanding of cultural change in Britain.

THE GEOGRAPHY OF CHANGE

Where to begin? The literary sources, that have at times baffled and disappointed many historians and archaeologists in attempts to define, for example, the invasion routes of Claudius' army, seem an obvious starting point. But we need to be clear what they represent. They do not create a total vision of the invasion but a fragmentary account that places an emphasis in their narratives on the change from a disunited series of kingdoms into the unified province of Britannia. The terminology used by Tacitus, Suetonius and Cassius Dio is instructive for our understanding of how Britain was viewed in terms of political geography. Tacitus (*Ann.* 2.24) refers to Roman contact with the *regules* (little kings) of southern Britain in his account of Germanicus' campaigns in Germany. This use of the diminutive would seem to have been quite deliberate for the discussion of the nature of the political leaders and the political formations associated with them.[7] These little kings ruled over *gentes* (peoples) identified, for example, by Tacitus through names such as the Iceni or the Silures.[8] Cassius Dio (60.20) informs his audience that the Britons were not free but under kings and divided into groups. The presence of clear *gentes* living in a number of towns or *oppida* is highlighted in the commemoration of the conquest of Britain. The account of the triumphal regalia awarded to Vespasian, during the governorship of Aulus Plautius and the presence of the emperor Claudius in Britain, mentioned the defeat of two *gentes*, the capture of twenty *oppida* (towns/hillforts) and the island of Vectis (Isle of Wight).[9] The celebration of Claudius' victory in Rome is also instructive; Claudius was saluted *imperator* several times 'contrary to precedent, for no one has received this title more than once for one and the same war'.[10] This suggests that rather than a general war against a people called 'the Britons' there was a series of wars between Rome and the various kings of the *gentes* of Britain, yet this series of wars was viewed as a single entity by geographical relation to the place of Britain, a fact acknowledged in the celebration of Claudius' victory on an annual basis in Rome through the re-enactment of the storming and *direptio* (sacking) of an *oppidum* as well as the surrender of the kings of the Britons.[11] It is clear that some kings/*gentes*, such as the Iceni, came to terms without any warfare taking place in AD 43. These were the eleven kings of the barbarian *gentes* mentioned on the arch set up by the senate and people of Rome to commemorate the victory in Britain.[12]

Warfare continued to occur after the initial period of conquest and was written up in a number of historical narratives that can be used to define some of the geographical changes taking place. In AD 48 an unnamed enemy attacked the fields of Rome's allies. A deliberate policy of disarmament of those living in the south of modern England was undertaken by Rome, in line with the *Lex Julia de Viis*. The opposition to this action came from the Iceni and those peoples identified as the *nationes* neighbouring that *gens*.[13] This grouping was defeated in battle and the Roman governor campaigned against the Ceagni taking the campaign of devastation and plundering across Britain to the Irish Sea. These events were accompanied by a brief revolt by the Brigantes, quelled by the execution of a number of leading men. However, the problem *gens* for the Roman governor was the Silures commanded by Caractacus, who could be neither intimidated nor converted by clemency. The only solution was to relocate the legionary fortress at Camulodunum to intimidate the Silures. The campaign against Caractacus resulted in triumphal insignia to be awarded to the governor and the appearance of Caractacus in Rome in front of Claudius and Agrippina.[14] The continuing resistance of the Silures was associated with the Roman view that their name and people should cease to exist. An achievement that was not carried out, but almost attained by Agricola later in his attack on the Ordovices.[15] We may compare the contemporary genocide of the Nasamones in Africa and the emperor's boast: 'I have forbidden the Nasamones to exist.'[16] This account of the actions of Rome contrasts the enemy or untrustworthy *gentes*, the Iceni, the Silures and the Ordovices, with the allies of Rome, the Brigantes, Cogidubnus and other allies to be imbued with a sense of legal obligation by the establishment of the *colonia* at Colchester. The variation in approach that allows for *civitates* to be handed over to Cogidubnus to ensure their loyalty through his kingship, the deployment of Roman military units to support certain kings and open hostility reveals a complex situation in Britain in which Rome had enemies as well as allies.[17] What is clear from the treatment of Rome's enemies in Britain by Suetonius Paulinus, a man said to be emulating Corbulo in the east, is that Rome's enemies were decimated by battle, plunder and famine from which they may not have recovered.[18] It is significant that in Ptolemy's lists of *poleis* in Britain (see below) the Iceni, the Silures and the Ordovices have only one or two places that were deemed to be *poleis*, whereas Rome's consistent allies the Brigantes have nine designated *poleis*. The defeated enemies of Rome, for example the Trinovantes, were subdued and humiliated, and it was only later through diplomacy and clemency by Agricola that Britain was pacified and hostages were sent to Rome's representatives.[19] The mixture of allies and enemies in Britain demonstrates the complexity of the historical situation of the early period of the province; this is a factor to be borne in mind when we discuss the production of towns and geography after the conquest.

The traditional picture of the development of towns in Roman Britain comes from Tacitus' account of the pacification of the province by his father-in-law Agricola.[20] In the winter after his second summer campaigning in the province, he encouraged those people 'living in isolation or ignorance' to build temples, *fora* and good houses. He also had the sons of the powerful educated and the toga was adopted. This much-discussed passage suggests that the transfer from a society based on warfare to one based on city life and peace was a deliberate policy of Agricola to pacify Britain. The emphasis on the promotion of urbanism and Roman culture during the winter after the campaigning season also appears in Dio,[21] and we might assume this to be an aspect of Roman government in areas in which there were *no* towns or civic structures. However, these actions of Agricola in AD 78 have come to dominate the modern literature and to create an illusion that what was being founded in all parts of Britain was a single *civitas-capital* for each of the peoples of southern Britain as well.

There is another history of town construction that depends on the interpretation of capital investment on the part of the Roman élite at the time of conquest. Claudius had given or loaned large sums of money to various prominent Britons after the invasion and other Romans had invested money into Britain – loans that were called in by the procurator in AD 62 – some fifteen to twenty years later.[22] Dio cites the loan of 40 million *sesterces* made by Seneca to the Britons made presumably after his return from exile in AD 49 as an example of the scale of credit/debt available to the Britons. What these loans and gifts of money were used for is less than certain. However, Cogidubnus utilised considerable funds for construction at Chichester.[23] *RIB* 91 refers to the construction of a temple to 'Neptune and Minerva for the protection of the divine house' by permission of Cogidubnus, but at the expense of a *collegium* of *fabri*. The presence of a *collegium* suggests an organisation for the burial of its members and by implication the concept that the members need not have close kin or dependants to undertake this action – that is, the craftsmen and labourers were not native to Chichester. The building of a temple by a *collegium* would imply that the members were wealthy and regularly employed. This in itself would suggest that a number of large projects were being undertaken by the members of the *collegium* within the region of Cogidubnus' authority as king.

The relationship between Rome and the monarchs of Britain prior to the arrival of Agricola can in part be revealed with reference to Queen Cartimandua of the Brigantes. She secured her position as a monarch dependent on Rome by handing over the fugitive Caractacus.[24] This event is hard to date, but the narrative of the triumph in Rome involves the figure of Agrippina as Claudius' wife, so should not be associated with the triumph of AD 44 but with another triumph after AD 48. What is clear, though, is that Cartimandua's wealth was derived from this action and when her kingdom was overthrown by the revolt of her husband in AD 69, cohorts of Roman soldiers re-established her power.[25]

These examples show Roman intervention prior to Agricola's first year in office through rewards to individual monarchs and kingdoms. However, we should not view this as a standard pattern across Britain. What is clear from the narrative of the *Agricola* (20–1) is that Tacitus is only referring to those peoples who were independent or hostile to Rome – who were suppressed by plundering raids to inspire fear, and clemency to cause their enemies to send hostages to Rome. Subsequently, these peoples were encouraged to found civic structures that had not previously existed. In contrast, the pattern of earlier investment into Britain was determined by the politics of friendship with Rome in the post-invasion period. These historical factors should underpin the explanation of the structure of Ptolemy's *Geography* that preserves earlier accounts of the number of towns associated with each *gens* in the AD 70s and 80s. The emphasis is on a lack of uniformity in the number of *poleis* associated with an individual *gens* and a pattern that contradicts the modern model of the *civitas-capital* with large rural hinterlands as standardised throughout the province of Britain.

PTOLEMY'S ROMAN BRITAIN

Ptolemy's listing of the geography of Roman Britain has been the source of some controversy and frustration, since the work does not tell the modern reader the answer to questions that Ptolemy never intended to be asked of the document.[26] This has led many authors to become dismissive of the evidence given. Rivet and Smith suggest that the geographer was only concerned with listing a quantity of cities rather than making a judgement of their right to be included on the overall assumption that there were – 'areas like northern Britain or free Germany, where no recognisable cities can have existed'.[27] The problem here is that Ptolemy may have viewed or defined the *polis* in a quite different way to that assumed by these modern authors. The listing of places is seen by Rivet and Smith to be a composition of earlier accounts with the addition of more up-to-date information on the location of various legions during the reign of Hadrian.[28] They see the account of the southern part of Britain to date to the AD 70s and the northern half to be constructed in the 80s. Rather than the developed province of the second century AD, we see here a description of Britain in the first century AD just over a generation after the Claudian invasion in AD 43. Hence, what we read here is an account of the developing province in which numerous towns are listed. This raises the subject of the recognition of urban forms in classical antiquity. We need to realise that a town or city need not have public buildings to be included in the category of *polis* and should not be used as a means to remove Ptolemy's evidence. After all, it is made clear by Pausanias in his discussion of the city of Panopeus (in modern Turkey), written in the second century AD, that a city need not include material remains of public architecture, but had to include a territory and headed a local league of cities.[29] Hence, when we read *polis*, we should view the place, whether a legionary fort or army camp, as the town in the region that has a role in local government and was a place of authority over an area and/or people. The fact that these very different forms (to our minds at least) of nucleated settlements were considered to have been *poleis* reveals much about the developing provincial structure and little about its final form.

Like most ancient geographers Ptolemy describes the coastline of any region by listing the places in order of a *periplus* around the coast. This he does for Roman Britain (2.2), but in contrast to his description of the coast of Italy, we find few towns mentioned apart from places called the Great Harbour above the Isle of Wight and another place called New Harbour; the other places listed are all physical features such as promontories or estuaries of rivers. He then goes on to list the peoples of Britain, their geographical position and the towns (*poleis*) associated with them. Here he includes places on the coast not mentioned in the earlier description of the shape of Britain. He includes legionary forts in his category of 'town'. This seems reasonable since the population of a legion was the size of a town even in the Mediterranean and these forts did exist within the territory of a people such as the Brigantes. Of the total of twenty-three peoples listed, the majority are associated with more than one town in its territory. This places a question mark over the ideal in the modern literature of a single capital for each of these peoples from which a vast territory was controlled as a standardised pattern for Roman Britain. This pattern is not unique to Britain and may be compared to Ptolemy's treatment (3.1) of the towns and peoples of central Italy. Like in Britain, there is considerable variation in the number of towns associated with a single group of people, from the Sabines with just one city to their neighbours the Tusci with over forty towns constituted as colonies or *municipia*. The variation in both Britain and Italy in the listing of peoples and their towns should be seen as deliberate and is contrasted with other regions in which the towns are simply listed separately from those of peoples in those provinces – notably in the cases of Noricum, Pannonia, Mauretania and Africa. A choice has been made in the creation of the list of towns for Britain that would have had a relevance to those that read the *Geography* in Rome and elsewhere in the Mediterranean. The peoples listed were seen as different from one another and the towns were places belonging to their specific *gens*. This would suggest, for example, that the Belgae had three towns: Isca, Aquae Calidae and Venta; whereas the Atrebates had a single town, Calleva. What we are seeing here is a variation in the pattern of recognition of what is a town and whether there was a political will or cash injection to develop one place as a town or an attempt to establish a number of places with this status. It should be noted that measurements and bearings in Ptolemy's text alongside the nature of their textual transmission create numerous geographical errors.[30] However, what is important is that the *gentes*/*civitates*/'tribes' are anchored with respect to their *poleis* as opposed to their actual geographical territory within Britain.[31] Therefore, the association between a *gens* and *polis* is secure; whereas the geographical position of the *gens* is far from accurate and the geographical location of some *poleis* is subject to error. Instead of a unified pattern of a *civitas* with a single town associated with a large territory, we find variation across Britain in terms of the number of *poleis* associated with each *gens* (Table 4.1 below) that is found also in Italy (Table 4.2 below).

We do not find in Ptolemy's version of Britain major differences in the overall pattern of towns and peoples. In Italy, in a few regions associated with major routes or in areas close to Rome, in other words those regions with a greater level of connectivity, there were significantly more towns than in the regions of Britain. But, elsewhere, for example in the Sabina and other regions isolated from the system of transport, a similar number of towns are found to those listed for Britain. There is a clear similarity between the geography of Britain and the Mediterranean core, but we can identify geographical difference between Britain and the North African and Danubian provinces.

Table 4.1 Britain: peoples and towns in Ptolemy's *Geography*

People	*Towns*	*Number of towns*
Novantae	Locophibia	2
	Rerigonium	
Selgovae	Carbantorigum	4
	Uxellum	
	Corda	
	Trimontium	
Otalini	Coria	3
	Alauna	
	Bremenium	
Vacomagi	Bannatia	4
	Tamia	
	Pinnata camp	
	Tuesis	
Venicones	Orrea	1
Taezali	Devana	1
Brigantes	Epiacum	9
	Vinovium	
	Caturactonium	
	Calatum	
	Isurium	
	Rigodunum	
	Olicana	
	Eburacum Legio VI victrix	
	Camulodunum	
Parisi	Petuaria	1
Ordovices	Mediolanum	2
	Brannogenium	
Cornavi	Deva Legio XX victrix	2
	Viroconium	
Coritani	Lindum	2
	Ratae	
Catuvellauni	Salinae	2
	Urolanium	
Iceni	Venta	1
Trinovantes	Camulodunum	1
Demetae	Luentinum	2
	Maridunum	
Silures	Bullaeum	1
Dobuni	Corinium	1
Atrebates	Caleva	1
Canti	Londinium	3
	Daruernum	
	Rutupiae	

continued . . .

Table 4.1 continued

People	*Towns*	*Number of towns*
Regni	Noeomagnus	1
Belgae	Iscalis	3
	Aquae calidae	
	Venta	
Durotiges	Dunium	1
Dumnonii	Voliba	4
	Uxella	
	Tamara	
	Isca Legio II Augusta	

Table 4.2 Central Italy: peoples and number of towns in Ptolemy's *Geography*

People	*Number of towns*
Gallia Togata	13
Tusci	41
Semnones	6
Piceni	12
Umbri	19
Sabini	1
Aequiculi	2
Marsi	2
Praetuti	2
Vestini	4
Marrucini	1
Latini	26
Peligni	3
Frentani	2
Caraceni	1
Samnites	8

THE STRUCTURE OF THE ROMAN PROVINCE: EXPRESSIONS OF MOBILITY

In the inscriptions from Britain, we receive glimpses into the integration of the island south of its frontier. Inscriptions record the deaths of people away from their home *civitates* or the building work on Hadrian's Wall conducted by people from a particular *civitas*. The roads were identified as a key element of the Roman project in Britain by Tacitus, alongside the encouragement of the adoption of Roman-style urban living. The roads were built through forced labour – a form of taxation, in contrast to the voluntary nature of town-building.[32] These roads reduced the temporal and physical distance and allowed for the measurement of distance between places; and gave a geographical unity to the province. The places found on the major routes serviced the travellers, whether soldiers, merchants or the provincial governor, and created or maintained the possibility of a Roman province that was a unified geographical entity at the macro-level. Here infrastructure needed for imperialism might be the key to understanding the geographical development of towns. Places on the route that soldiers took could have expected to have benefited from those soldiers' expenditure en route. Similarly, the provision of services at *mansiones* integrated these places into a wider Roman culture via the presence of travellers stopping, resting and ultimately spending money at these locations. Equally, the movement of credit in the early years of the province was followed closely by that of foreign workers.

The Antonine Itineraries provide us with an account of the major routes and places on those routes for not only Britain, but the whole of the empire (see Salway above). It would appear that the document is a compilation of earlier itineraries and incorporates these to create an account of the geography of the whole empire.[33] This allows for a comparison to be made across the empire between provinces. The individual itineraries list places in consecutive order along key routes (Figures 4.1–4.15). The itineraries intersect and build up into an account of the geography of places and the distances between places for the provinces and Italy (coverage can be patchy, for example, in western Gaul, but this is not the case for Britain). There is a natural emphasis in these texts to view space as linear, connecting places in between, so that they could have been utilised either in forward or reverse order to create the necessary route. The

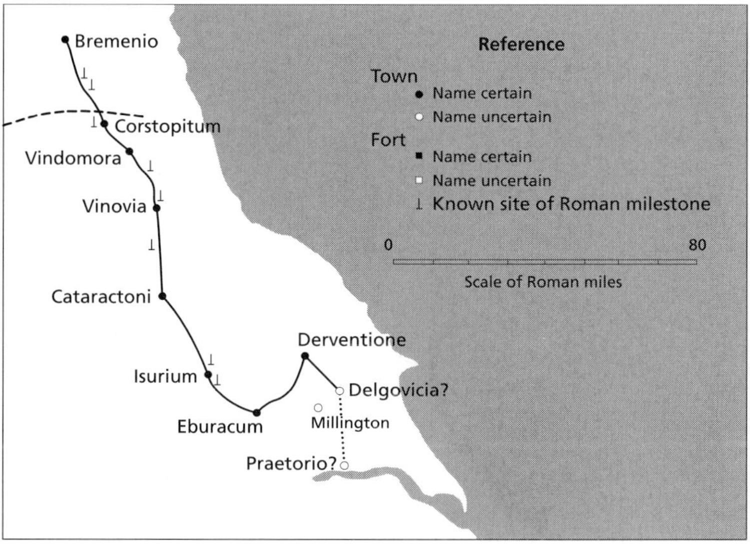

Figure 4.1 The course of *iter* I (after Rivet and Smith)

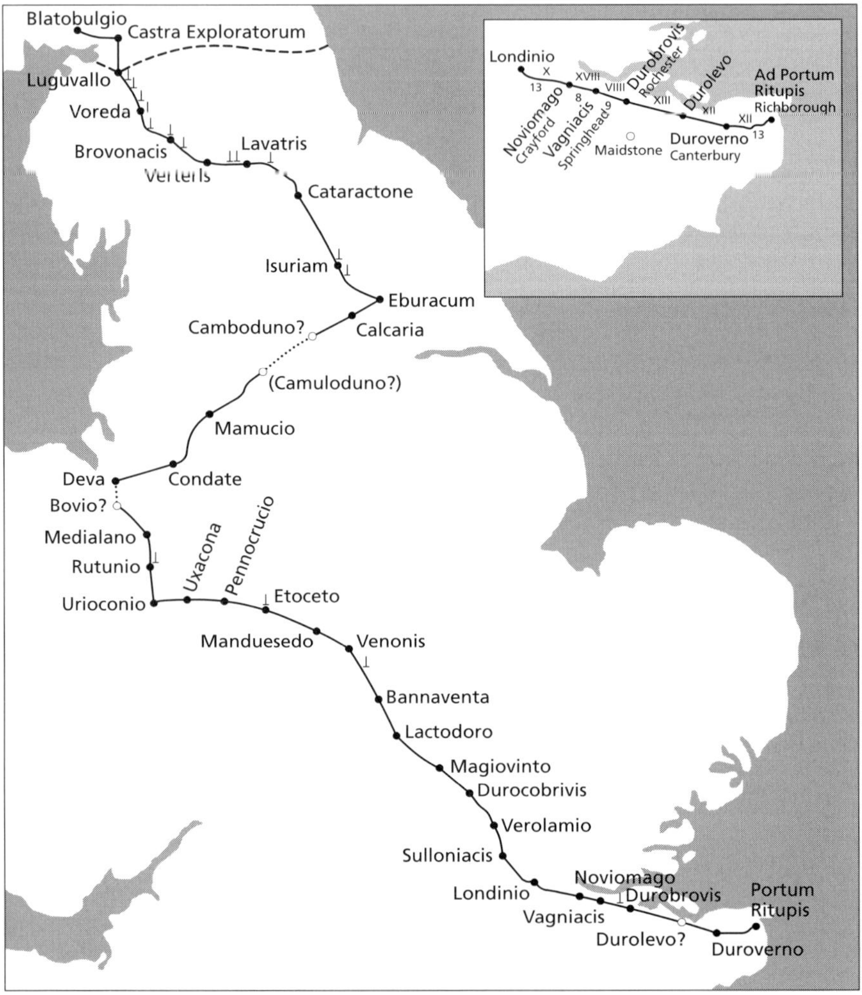

Figure 4.2 The course of *iter* II (after Rivet and Smith)

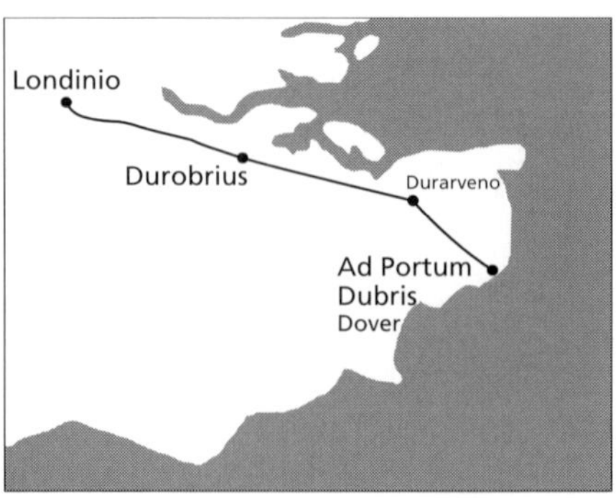

Figure 4.3 The course of *iter* III (after Rivet and Smith)

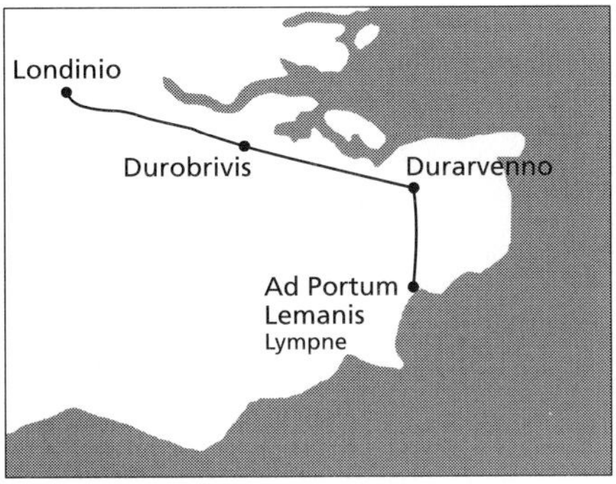

Figure 4.4 The course of *iter* IV (after Rivet and Smith)

Figure 4.5 The course of *iter* V (after Rivet and Smith)

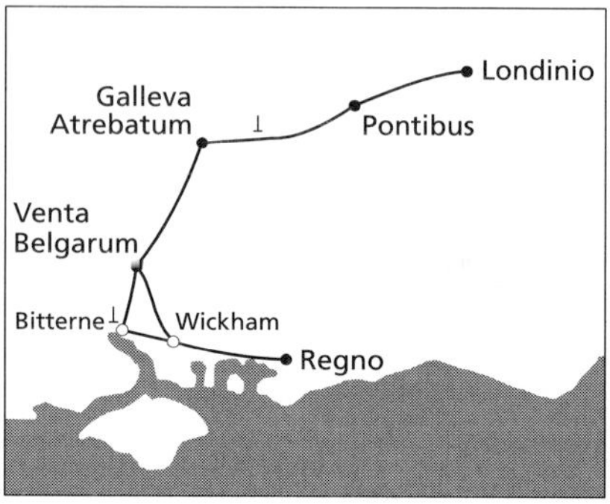

Figure 4.6 (top left) The course of *iter* VI (after Rivet and Smith)

Figure 4.7 (above) The course of *iter* VII (after Rivet and Smith)

Figure 4.8 (bottom left) The course of *iter* VIII (after Rivet and Smith)

Figure 4.9 (below) The course of *iter* IX (after Rivet and Smith)

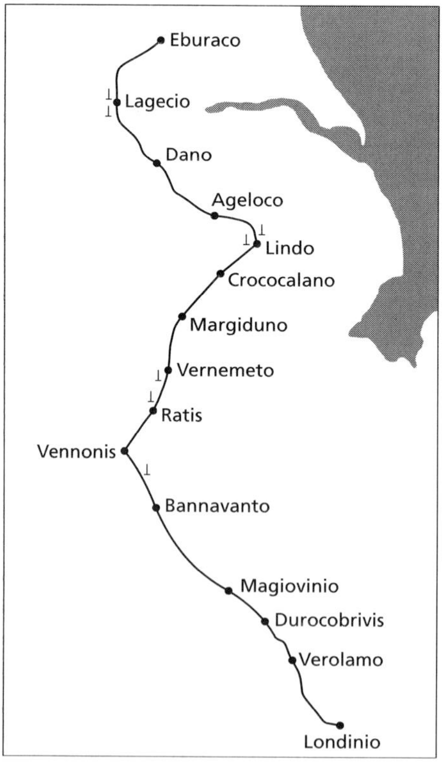

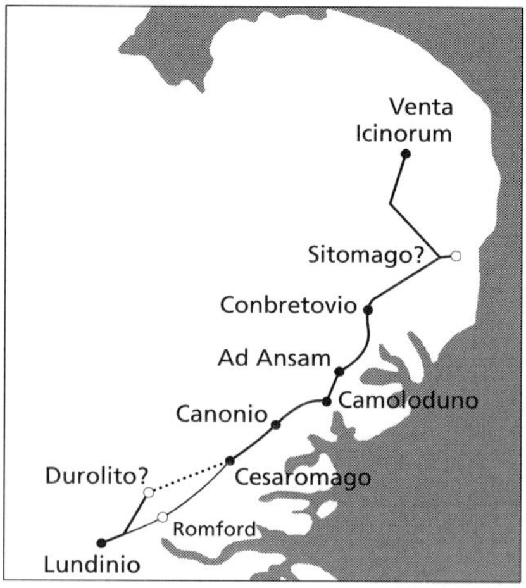

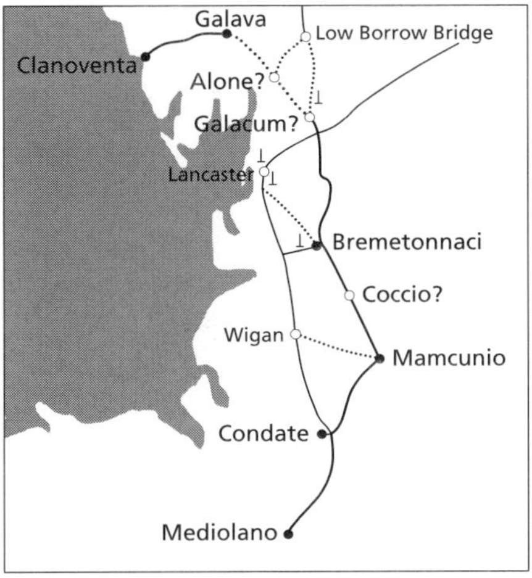

Figure 4.10 The course of *iter* X (after Rivet and Smith)

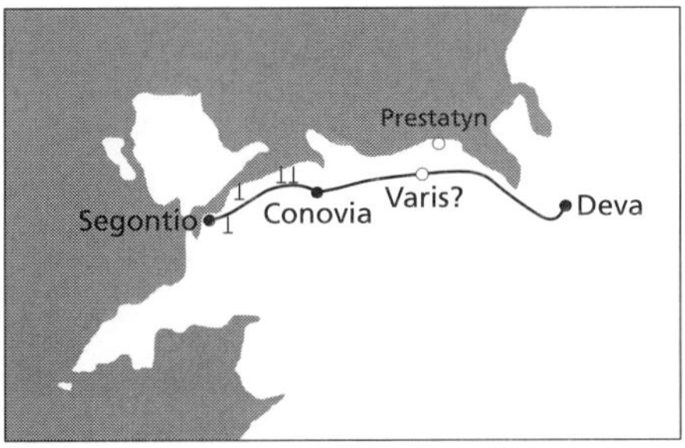

Figure 4.11 The course of *iter* XI (after Rivet and Smith)

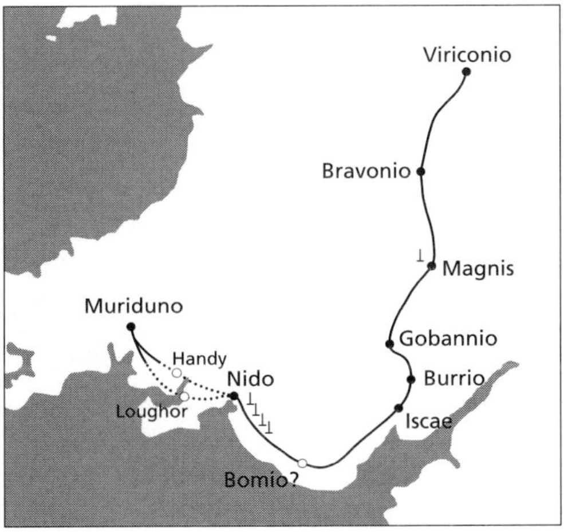

Figure 4.12 The course of *iter* XII (after Rivet and Smith)

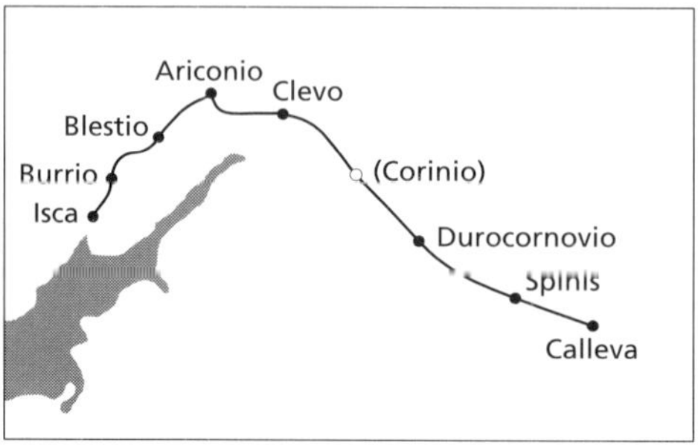

Figure 4.13 The course of *iter* XIII (after Rivet and Smith)

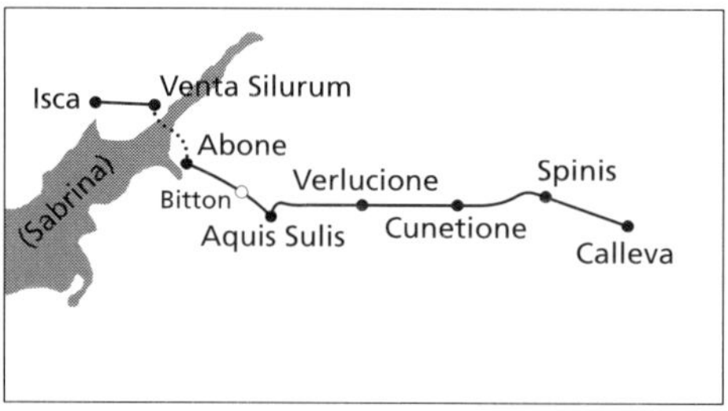

Figure 4.14 The course of *iter* XIV (after Rivet and Smith)

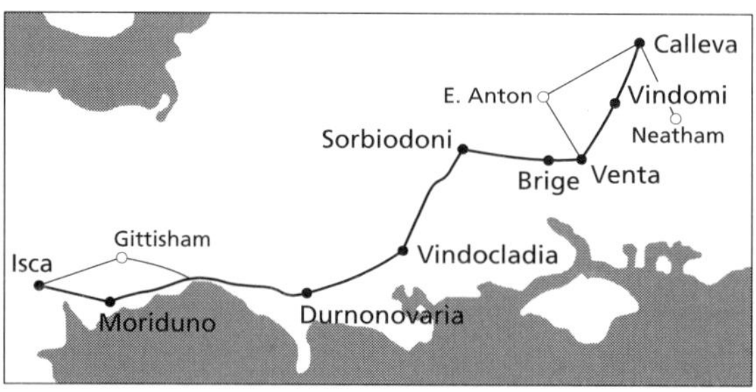

Figure 4.15 The course of *iter* XV (after Rivet and Smith)

journeys listed should be viewed not individually but as an interrelated network of routes that intersected at certain points or key places. The stress in the itineraries on the distance between places in both peripheral provinces and the Mediterranean core allows for comparison of inter-centre distances or the spacing of places in a number of locations across the empire.

The Antonine Itineraries for Britain begin north of Hadrian's Wall with the first running down the eastern side of the province to Praetorium on the Humber estuary. The second itinerary runs down the country from Carlisle via London to Richborough in Kent. Then are added the third and fourth itineraries from London to the other Channel ports. The fifth itinerary returns to Hadrian's Wall, but this time to the eastern end – the starting point of the first itinerary. These itineraries create the link between the south and the north, between the points of arrival into the province and the edge of the province viewed as Hadrian's Wall. There follow four itineraries connecting London to Lincoln, Chichester, York and Venta Icinorum – all either key legionary fortresses or the towns of early client kings. The tenth itinerary from Ravenglass to Whitchurch seems to be odd at first sight since it does not fit the pattern of connecting Channel ports to Hadrian's Wall or London to key places of either legionary settlement or client kings in the province. However, we can find Whitchurch in the second itinerary from Carlisle to Richborough and it is the intersection of the tenth itinerary with the second itinerary that integrates the two to create additional journeys. Whitchurch in this case acts rather like a hub airport today from which other journeys are made after the initial journey. The eleventh itinerary is also concerned with the Ordovices and runs from Chester to Caernarvon and can be seen as another spur from the second itinerary from Carlisle to Richborough. Similarly the next itinerary from Camarthen to Wroxeter interlinks with this long north–south itinerary. The last three itineraries detail routes from another hub – Silchester. The first two lead west to Caerleon and the final one to Exeter to incorporate the geography of south-west Britain.

The starting and finishing points of the itineraries are instructive as they relate to places of particular significance in the landscape of Britain: the coastal points of contact – the Humber, Ravenglass, Richborough, Dover, Lypne, the eastern and western ends of Hadrian's Wall; the provincial capital London; the legionary bases; the towns of significant client kingdoms of the early conquest period – Chichester, Silchester and Caistor by Norwich; and regions that were on the edge of the province – Wales, Exeter and Hadrian's Wall. The spacing on all the journeys listed for the most part falls below thirty miles with an average spacing of between thirteen and twenty-three miles. These are all distances that could be easily travelled on foot in a single day.[34] In other words, no point of Britain listed in the itineraries was isolated or inaccessible to the traveller. This in itself should be seen as a major factor in the creation of the geographical area known as Britannia from a disunited set of kingdoms or the process often termed 'Romanisation'.

A key question is whether the structure of the itineraries for Britain reveals a geography that is distinct from the itineraries of other parts of the Roman empire. When compared to the data for Italy, there is no significant difference between these itineraries in the range of miles between each stage, or the average spacing of places along the routes.[35] This is shown in Figure 4.16, where the inter-centre distances in the Antonine Itineraries for Italy and Britain are plotted against each other. The sample size is nearly the same, 163 entries for Italy and 160 entries for Britain. The spacing of places in Italy reveals a median in the upper range of sixteen to twenty miles; whereas the median spacing in Britain

Table 4.3 The fifteen Antonine Itineraries from Britain

Iter		*Number of stages*	*Total distance*	*Range*	*Average spacing*
1	Vallum–Praetorio	9	156	7–25	17
2	Vallum–Portum Ritupis	36	381	9–24	14
3	Londinio–Portum Dubris	3	66	14–27	22
4	Londinio–Portum Lemanis	3	68	16–30	23
5	Londinio–Vallum	19	442	16–35	23
6	Londinio–Lindo	13	156	7–21	13
7	Regno–Londinio	5	96	10–27	19
8	Eburaco–Londinio	14	227	12–28	16
9	Venta Icinorum–Londinio	8	127	6–32	16
10	Clanoventa–Mediolano	8	110	12–27	18
11	Segontio–Devam	3	74	23–32	25
12	Muridono–Viroconorum	17	186	8–36	11
13	Isca–Calleva 1	8	109	9–15	13
14	Isca–Calleva 2	8	103	6–15	13
15	Calleva–Isca	8	136	8–36	17

was well within the eleven–fifteen-mile grouping. The overall pattern for the two regions – one at the centre and the other at the periphery of the empire – reveals a fundamental geographical difference that we would not predict. The spacing of places on the routes in the Antonine Itineraries is much closer for Britain than Italy, although the overall range is similar. The shorter spacing of places and towns found in the British section of the Antonine Itineraries might be accounted for in terms of road technology. Whereas roads in Italy had been paved with durable materials, the roads of Britain for the most part were seldom paved but formed of compacted gravel.[36] Hence, the differential in type of road surface, and for that matter maintenance at a distance from the political centre, could have required the spacing of stopping points on the roads of Britain to have been more frequent.

To make the same comparison with Gaul (Figure 4.17) we find once again a notable difference in the frequency of intervals between places. The grouping spaced at a distance of between eleven and fifteen miles was far greater in Britain than Gaul. The pattern in Gaul is in fact closer to that of Italy but with an overall emphasis on rather shorter distances. This suggests that at the centre of the empire distances between towns or stopping points were greater than they were in the north-west provinces. This view is backed up by the comparison of distances in Britain with those found in Spain and Asia (Figures 4.18 and 4.19). These two provinces as listed in the Antonine Itineraries conform to the Italian pattern of distance with a greater emphasis on inter-centre distances in the range of sixteen to twenty-five miles. We might see the patterning of inter-centre distances in the Mediterranean as the developed form of settlement

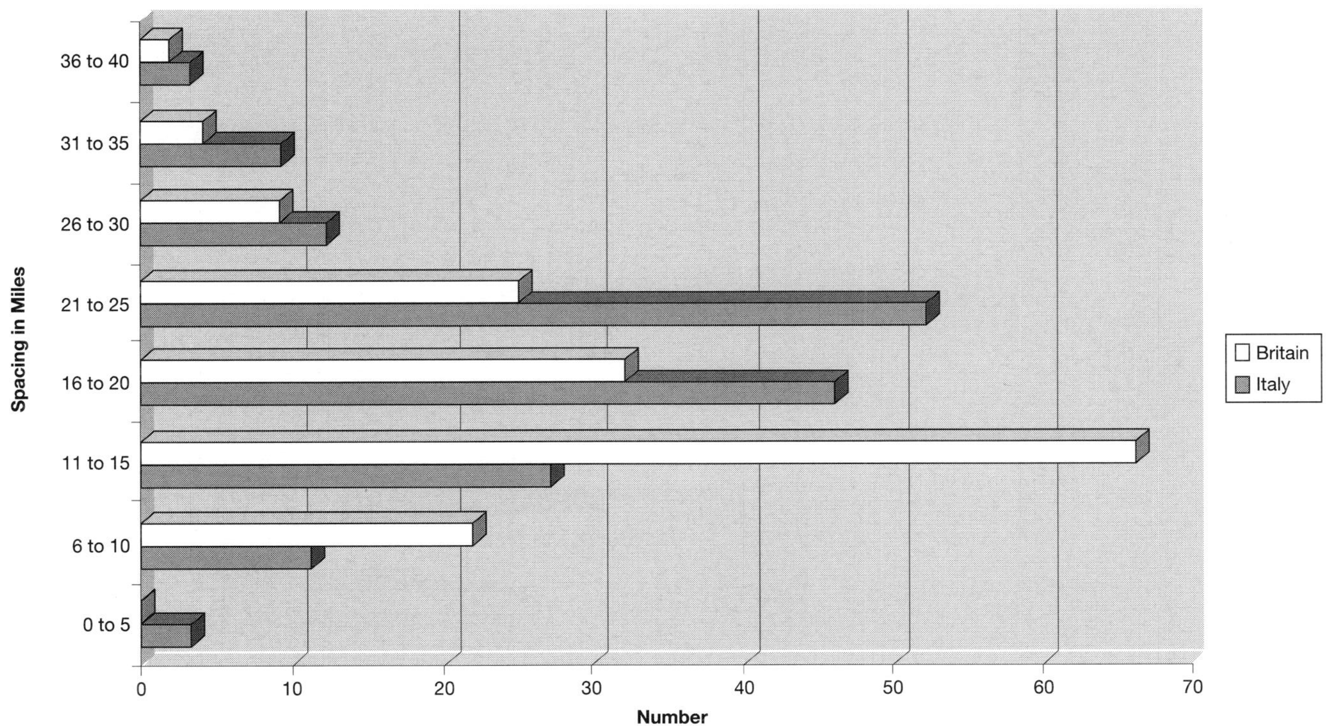

Figure 4.16 Comparison of inter-centre distances in the Antonine Itineraries I: Italy and Britain

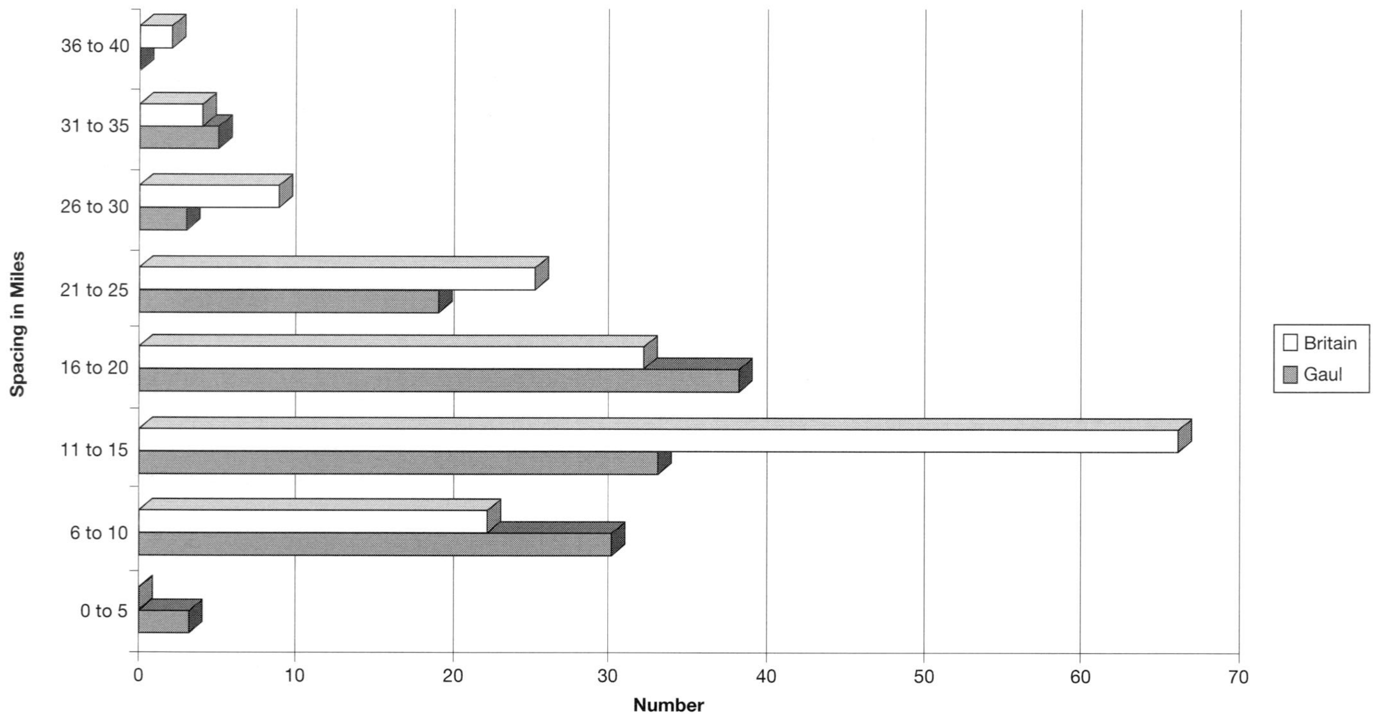

Figure 4.17 Comparison of inter-centre distances in the Antonine Itineraries II: Britain and Gaul

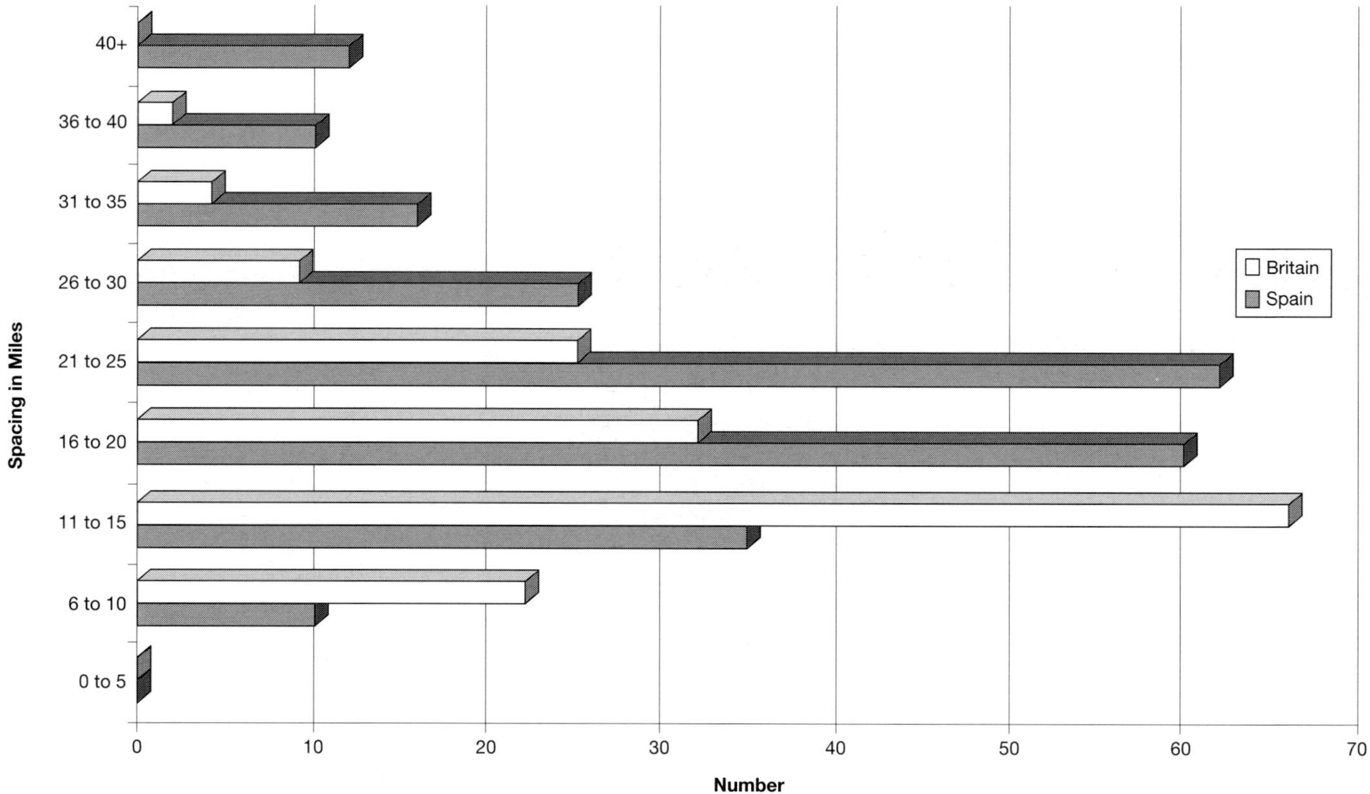

Figure 4.18 Comparison of inter-centre distances in the Antonine Itineraries III: Britain and Spain

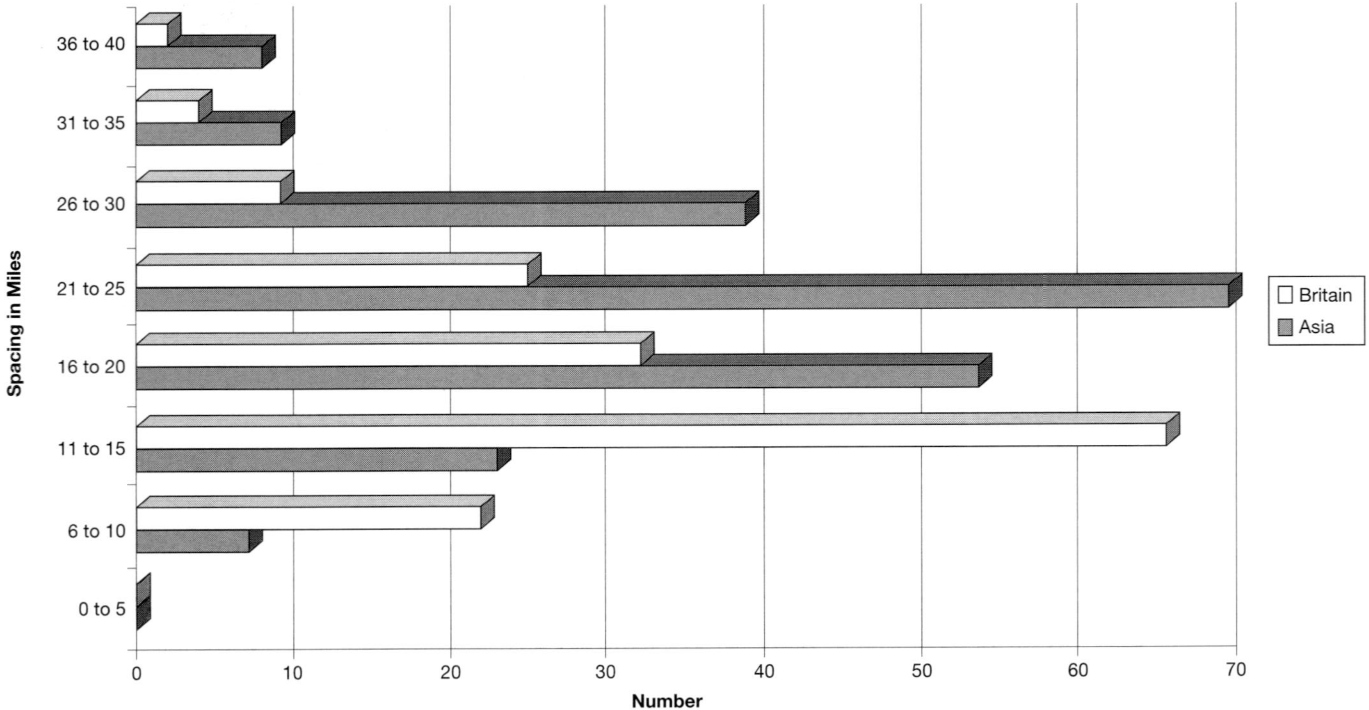

Figure 4.19 Comparison of inter-centre distances in the Antonine Itineraries IV: Britain and Asia

geography, whereas the emphasis on a larger number of places closer together is the less developed form. This contradicts the studies of the patterning of towns found in Pliny the Elder's *Natural History* and Strabo's *Geography*. The work of Bekker-Nielsen based on the study of inter-centre distances in Gaul and Italy found in Pliny pointed to an overall increase of distances between places from central Italy through to northern Gaul.[37] Why the distances found in the Antonine Itineraries do not conform to this pattern may be because all the places identified are on designated routes with an interconnection with their nearest neighbours along the road, rather than across country as well. However, this does not explain the overall pattern of lower inter-centre distances in Britain than in Italy. Perhaps the need for a greater number of stopping places reflects less development in the infrastructure of travel in Britain. This required that the places in which travellers could stop was at a greater frequency – a factor that would affect the overall concentration of resources and the development of towns in larger, more centralised nucleated settlements.

In terms of the development of towns and nucleated settlements, the need for traffic to stop more frequently would not have promoted the centralising of resources to a few centres, but would have promoted a greater number of small nucleated settlements or simple stopping points. The latter have been recognised in some cases archaeologically at, for example, Catterick, Wanborough and Wall by the characteristic building type seen to have been a *mansio*.[38] These *mansiones* have been seen as an impetus for the creation of nucleated formations that are associated with ribbon development along the routes of travel around the *mansio* site.[39] However, it is for the most part unclear whether the *mansio* was the impetus to such development or whether the *mansio* was placed in these locations in response to urban development, but it should be noted that most of the small towns of Roman Britain lie on transport routes and a connection between urban development and long-distance land transportation seems convincing.[40] Our picture of the distribution of such small towns is patchy and dependent on outside forces of the present rather than simple survival rates reflecting the past. Many towns mentioned by Ptolemy or the Antonine Itineraries simply do not appear. However, the link between transport and nucleation is proven by Burnham and Wacher, even if these nucleated sites do not display the monumental trappings of a Roman city (or *civitas-capital*). They should nevertheless be regarded as towns since there was no requirement for towns in Britain or the Mediterranean to have public buildings in order to be referred to as urban structures or *poleis*.[41] The settlement pattern for Britain along the major transport routes found in the Antonine Itineraries displays a greater density of named places, but archaeologically these need not be more than a simple *mansio* with its associated settlement and supply base of animals.[42] The multiplicity of such towns reduces the significance of a larger central place for every *gens* of Roman Britain. However, some sites developed a greater importance and are indicated as such by their position in the Antonine Itineraries as starting or end points for a number of individual routes. London to the north of the Thames with its link to the Channel ports was not only the provincial capital but also the crossing point on the Thames itself leading to the north. In the south Silchester is the key nodal point for three itineraries to the south and west.

THE HISTORIOGRAPHY OF URBANISATION IN BRITAIN – FROM FACT TO MODERN FICTION

Much emphasis has been placed on the role of towns in the discussion of the Roman impact on Britain. The literature has produced a model of the promotion of town-building and the centralisation of power in the town or *civitas-capital*. This emphasis on the town and its region found in the Mediterranean has been challenged by Horden and Purcell.[43] They put forward a view that the town is not the motor for cultural change, but an arena, an address and an architectonic agglomeration that was fundamentally affected by the mobility or flow of people, goods and capital. The question for us is how should we view the town in Britain in relationship to the mobile culture that we associate with the Roman empire?

There has often been an expectation that Rome was aiming to create a version of the Mediterranean city-state formation in Britain that would of necessity cause or stimulate changes to the existing geographical and political organisation of the new province.[44] There is assumed to have been some form of hierarchy of towns: provincial capital, *coloniae*, *municipia* and *vici* as found in a number of legal texts.[45] I shall argue that this emphasis on the development of a legalistic hierarchy confines the debate and does not represent the nomenclature used – *municipium* occurs once only in connection with Verulamium and Boudicca in a tripartite literary flourish that includes the destruction of London the provincial capital and the *colonia* of Camulodunum.[46] What Tacitus is creating is an image of devastation that includes mention of these three different types of town for a Mediterranean-based audience. It is worth remembering that peoples across the empire adapted their extant political systems to utilise Roman legal or political terminology quite differently from those found in existing colonial or municipal charters from Italy and Spain; for example, an inscription from Brough records that an Aedile of a *Pagus* erected the proscenium of the theatre dedicated to Antoninus Pius and the *Numini Augusti*.[47] The development of a precise equation between a *civitas-capital* and a named people in the modern literature is based on very few specific examples.[48] Much evidence is ignored in favour of the modern model of the single *civitas-capital* based on very slender epigraphic evidence and extrapolation from the later compilation of itineraries known as the Ravenna cosmography.[49] This model has also influenced the discussion of 'Romanisation' of Britain.[50] There is a distinction in the literature between the settlements that are named as *coloniae*, *municipia*, provincial capitals, *civitas-capitals* and *vici*. The evidence for these formations is patchy: the *coloniae* are well represented in textual sources; in contrast we know positivistically of only one *municipium* from the province.[51] For our purposes here, London as a provincial capital seems very different to all other urban formations and will be left to one side. The last two categories – *civitas-capitals* and *vici* – are at times related to each other, with the *vici* being transformed or evolving into a *civitas-capital* at a later date.[52] Even the more revisionist accounts of Roman Britain adopt a model of town development in relationship to a territory in the manner of a Mediterranean-style city-state.[53] This model of each region having a clearly defined capital with no real rivals has been developed over the twentieth century to create a Roman aim to create a uniform political geography of Britain with little variation on the theme of a centralised town governing a vast rural territory – some ten times larger than the rural territories of towns in the Mediterranean.[54]

The historiography of this model of the political geography of Roman Britain needs to be examined to understand what its basis is in terms of actual evidence, and how modern interpretation has transformed this evidence to create a uniform Roman policy of urbanisation and development that need not have existed in antiquity. The origin for the idea of a hierarchy of towns which culminated locally in a *civitas-capital* was developed by Francis Haverfield in the early part of the twentieth century. He observed that Caerwent (Venta), a town known from an inscription to have been associated with an *ordo* of the *res publica civitatis Silurum* in the early third century AD, was listed in the Ravenna Cosmography as Ventaslurum; in other words as Venta of the Silures.[55] This addition of a tribal name to a town name was followed up with reference to other towns with two names: Venta Velgarom, Venta Cenomum, Rate Corion, Utriconion Cornoviorum, Duro Averno Cantiacorum, Cironium Dobunorum and Caleba Arbatium. All of these towns were viewed as tribal capitals of the province of Britain and this argument was elaborated by Stevens to identify further examples. However, there is a problem in the methodology here.[56] The way the towns in the Ravenna Cosmography have been compiled from numerous earlier lists is haphazard – for example, listing *gentes* (peoples) as places.[57] Also, there are within the Ravenna Cosmography a number of places with the same name: for example, Alauna and another Alauna with the *cognomen* Silva. This is not a unique phenomenon to Britain and needs to be explained with reference to well-documented examples from elsewhere in the empire. In Pliny's list of the towns of Roman Italy, placenames that are repeated were distinguished by an additional name – for example, Foroiulienses cognomine Transpadani and Foroiulienses cognomine Concupienses – as a means of precise identification.[58] In the same way, in the Ravenna Cosmography, we find Venta Velgarom, Venta Cenonum and Ventaslurum. No doubt these towns had the same name Venta and in addition were given a *cognomen* referring to the local *gens* to distinguish them from each other. Other towns that appear with a *cognomen* that refers to a *gens* within the Ravenna Cosmography follow this pattern also.

The notion of *cognomina* for towns of the same name removes any significance from the double-named towns found in the Ravenna Cosmography, which have been seen as the *civitas-capitals* of the province. What we have instead is a list of *civitates* in the late Roman period drawn from earlier lists; interestingly, forts (and *gentes*) are included as well as towns. This should be read in the context of the observation by Abbott and Johnson that the word *civitas* after AD 312 came to supplant all other terms for towns, including *coloniae* and *municipia*, and refers to the basis for taxation within a region of the empire.[59] Hence, we find people listed in inscriptions as *cives* of places far from where the inscription was set up,[60] and *civitatis* tends to be utilised in a similar way for actions taken by a group or paid for by a *civitats* far away from its location: for example, on Hadrian's Wall.[61] In no way do these inscriptions support the notion of each people with its own *civitas-capital*.

ROMAN BRITAIN: MOBILITY AND CULTURAL CHANGE

Our re-examination of the historical and geographical sources referring to Britain in antiquity has revealed a number of features that would have affected the place of travel within the culture of the province. The examination of Ptolemy's *Geography* and the Antonine Itineraries demonstrated that the structure of Britain's geography was similar to that of the core areas of the Mediterranean. Where differences were identified these were with the North African and Danubian provinces, in other words places where there were *gentes* without *poleis*. It is clear that what we see in the geographical sources is a developing provincial structure that had been given an impetus to do so by the investment or loan of money shortly after the conquest. It is impossible to know for how long this investment continued. However, the calling in of loans in the 60s AD clearly affected the attitude of the Britons to the Roman project in the island. In terms of connectivity, it has been demonstrated that Britain was not significantly different from the Mediterranean and that the creation of long-distance routes, listed in the Antonine Itineraries, did create the possibility of greater mobility within the island. Obviously, there is a major difference between places on these major routes and those isolated from them. In the discussion of towns, it was necessary to expose the weakness of the single *civitas-capital* model that has dominated the literature and to put forward an argument for an association between any number of towns and a specific *gens*. This is not to say that some of these towns could be larger than others, but to stress that the towns were associated with a specific identity of a people, that also can feature in inscriptions outside of the area associated with that identity. It is significant that the commemoration of these identities is almost exclusively made beyond the locale of a *gens*. We might say that the cultural change associated with the creation of the province of Britannia and a general promotion of mobility created a need to specify a person's identity outside of the place of their birth. The geography of unification of Britain by Rome may have reinforced or created these local or regional identities. However, we should be clear that these regional identities were not fixed and were subject to negotiation and change. Travel beyond Britain may have changed the person's perception of their identity, for example, from Brigantian to generically British.[62]

The view that the northernmost province of the Roman empire was different from the core areas of the Mediterranean has been questioned. This is done by altering our conceptual viewpoint. If we do not regard monumentalism within towns as the defining feature of Romanness and a sense of belonging to Rome, we can begin to understand the culture of the Roman empire in a different way. The lead is given by Horden and Purcell in *The Corrupting Sea* via a view that emphasises the fluidity of the movement of people, goods and capital through a network of towns. This has been shown to be a major factor in the Mediterranean and I think we may argue that it holds true for Britain as well. The difficulty is to identify the movement of capital. In a way, the monuments of towns display the concentration of expenditure in one place at a particular point in time, but that is not the whole story of urbanism. Other flows, such as the system of credit and debt that had its origins in the Claudian period in the imperial court and its immediate circle, are factors in the creation of monumentalism and expenditure within centralised locations. These flows of credit and global finance naturally followed lines of friendship rather than enmity and we should not underestimate the role of politics in the development of towns.

However, there is the issue of choice of how to spend money derived from Rome in the Claudian and Flavian periods. If we view the native élite as sophisticated and capable of reading the system of signs and signification associated with Rome and its empire, we can begin to realise that becoming a province or being Roman was something that was understood.[63] The major change can be seen to have been the creation of long-distance roads to promote mobility and the unity of a geographical area that had been ruled by *regules* (little kings). The replacement of the latter by governors and procurators would have provided direct contact with a Mediterranean culture. Other associated flows of people in the form of those in the army, the small-scale traders and craftsmen from beyond Britain would also create a system of contacts of a limited but of quantitatively significant number. Across the island the frequency of such contacts would vary, as would the connectivity between the various parts of Britain.

If Romanisation is to mean anything, it should be explained with reference to a change in the mobility and connectivity of the island of Britain. The system of towns and roads that were built and are referred to in our sources created a Mediterranean pattern of interconnection that was focused on travel from the south-east to the northern and the western extremities – a structure that reflected the politics of conquest and the need to garrison the province with an army. Not surprisingly, changes to the pattern or flow within a culture produce hostile reactions, sometimes termed resistance. However, it needs to be pointed out that these were of limited duration and did not produce long-term changes to the new structure of the Roman province of Britannia.

NOTES

1 An earier version of this chapter was presented as a paper at the Roman Archaeology conference at Durham in 1999. I would like to thank the participants and others who commented on the paper. I would like to thank my colleagues at the University of Reading for support for my research, but I owe a particular debt of thanks to John Creighton, who has discussed his view of cultural change in Britain and numerous related topics over the last few years. These discussions sparked my curiosity in the evidence for mobility and geography in Britain and resulted in this chapter. Naturally any errors or misconceptions expressed here are my own.
2 Horden and Purcell (2000).
3 Ibid. 9–25.
4 Laurence (1999).
5 Millett (1990) continues to be the most effective account of this process.
6 See, for example, papers in Mattingly (1997).
7 Analysis of Iron Age coinage and literary sources has recently produced a new history of the Kings of Britain – Creighton (2000).
8 Tac., *Ann.*12.31–2.
9 Suet., *Vesp.* 4. The language is traditional in that the burning of *vici* (villages) and fields was an assumed part of every military action; whereas the capture of towns was seen as glorious and out of the ordinary. See Tarpin (1999) 288; cf. Livy 10.17.2.
10 Dio 60.21.
11 Suet., *Claud.* 21.6; Dio 60.22; on the nature of *direptio* see Ziolkowski (1995) 69–91.
12 Tac., *Ann.* 12.31 for literary account; *CIL* 6.920 for the inscription from the victory arch in Rome.
13 Tac., *Ann.* 12.31.
14 Tac., *Ann.* 12.32–9.
15 Tac., *Agr.* 18.
16 Dio 68.6.
17 Tac., *Agr.* 14, *Ann.* 12.40.
18 Tac., *Ann.* 14.29–39.
19 Tac., *Ann.* 14.31, *Agr.* 20.
20 Tac., *Agr.* 21.
21 Dio 56.18 on the development of towns in Germany, a text discussed by Jones (1984) 75–91.
22 Dio 62.2.
23 Bogaers (1979) 243–54; Barrett (1979) 203–26.
24 Tac., *Ann.* 12.36–8.

25 Tac., *Hist.* 3.45.
26 For discussion of the nature of ancient geography see Nicolet (1991).
27 Rivet and Smith (1979) 105.
28 Ibid. 114–15.
29 Pausanias 10.4; Lomas (1998) 67 suggests that this passage indicates that by the second century AD the status of city/*polis* was judged according to the extent of monumentalisation. However, the passage also shows that monumentalism was not essential for the term *polis* to be applied to a place. In the context of Ptolemy and the slightly earlier date of his list, we could suggest that the emphasis on the structures of monumentalism need not have been the determining factor in the definition of *poleis* and their inclusion into his list.
30 Rivet and Smith (1979) 118–20, 129–31.
31 Ibid. 121.
32 Tac., *Agr.* 31; compare with 21.
33 Rivet and Smith (1979) 150–4. The exact dating of the Antonine Itineraries is fraught with difficulty; they are certainly later than the second century AD.
34 See Laurence (1999) 81–2 on speed of travel.
35 On the Italian evidence, see ibid. 84–93.
36 Ibid. 58–77.
37 Bekker-Nielsen (1989).
38 Burnham and Wacher (1990) 5.
39 Ibid. 23–7.
40 Ibid. 44.
41 As is made clear by Pausanias 10.4.
42 Compare epigraphic evidence for the structure of a *mansio* in Asia Minor discussed by Mitchell (1976).
43 Horden and Purcell (2000) 89–100.
44 Millett (1990) 65–8.
45 Wacher (1974) 17.
46 Tac., *Ann.* 14.3.
47 *RIB* 707; for comparison see Lomas (1995) 107–20 on the Greek cities of southern Italy.
48 For example, *RIB* 311.
49 Wacher (1995) 21–35.
50 Burnham (1979) 256.
51 Wacher (1995) 18.
52 Ibid. 17; Jones (1984).
53 Millett (1990) 65.
54 Haverfield (1912) 57–8; Wacher (1995) 17–35; Woolf (1998) 111–12.
55 Haverfield (1924) 194, based on *RIB* 311.15.
56 Stevens (1937) 193–203; and (1951) 188–92.
57 Rivet and Smith (1979) 194–5.
58 Pliny, *NH* 3.131, 3.113, respectively.
59 Abbott and Johnson (1926) 192 and 39–54.
60 *RIB* 621, 678, 693, 1593, 1594.
61 *RIB* 1672, 1673, 1843.
62 For comparison see Marshall (1998) 49–63, on the role of the ambiguity of a person's identity according to where and with reference to whom they are

situated. Also note the role of identity expressed by gladiators, a group who were inevitably away from their place of birth: see Hope (1998) 179–95.

63 Creighton (2000) presents a powerful argument from the imagery on Iron Age coinage to suggest that Britain became Roman after Caesar's conquest and that the client kings of Britain were active participants in the use of classical imagery. The extent to which this élite pattern is also reflected in more mundane aspects of material culture has yet to be shown.

5

TRANSPORT AND COMMUNICATION IN THE ROMAN STATE

The cursus publicus

Anne Kolb

The *cursus publicus* was a peculiarly Roman government institution which has been widely misunderstood. By making possible government communication, travel by government officials and the transport of certain freight goods, the *cursus publicus* played a vital role in securing and exercising Roman rule throughout the empire. The institution today still is termed 'post office' or 'government post' by many, even though its function and organisational structure were quite different from modern postal systems.[1] It is well known that the *cursus publicus* did not carry out appointed rounds, nor was it universally available.[2] This chapter seeks to illustrate more of such differences, in order to show that the *cursus publicus* was not a delivery service like a post office, but rather an infrastructure for use by state officials.

CREATION AND OPERATION OF THE CURSUS PUBLICUS

Although the public transportation system known to us as the *cursus publicus* is not mentioned by name in sources prior to the fourth century, we know from a chapter in the *Life of Augustus* by Suetonius that it was Augustus himself who established it.[3] The fundamental principles upon which its operations were based, under the principate as well as under the late empire, were laid down at that time. It seems justifiable, therefore, to extend our use of the fourth-century term to the earlier organisation as well.

About the creation of the *cursus publicus* Suetonius (*Aug.* 49.3) noted:

> So that he [Augustus] quickly and easily could receive reports of events

> in every province, he stationed first young men, and later carriages, at points along the military roads. The latter system proved itself more advantageous, because then the same courier could deliver the message to its destination and, if necessary, also personally be questioned.

The text states clearly that Augustus' primary concern was the transportation of important messages. Suetonius describes two methods by which messages were transmitted. At first, young men from local communities were posted at points along the *viae militares*, apparently as a sort of relay system of runners, who passed messages at each station from one to the next. As Suetonius notes, this system had the disadvantage that the courier had no personal knowledge of his messages' content and, therefore, could neither answer questions nor provide supplementary information about conditions or events at the place where the message originated. The system was later changed to include vehicles that could be exchanged at the staging posts. Under the revised system, the same messenger could travel from his point of departure directly to his destination.

As Suetonius recorded, Augustus intended that only vital official dispatches would be carried. The first system of runners was no more than a courier service. The second system, however, which remained in effect throughout the later period, could do more than just deliver official mail. The *vehicula* – that is to say pack animals, boats or, most specifically, wagons – could carry not only the couriers and their messages, but also other persons as well as a limited amount of baggage or other freight. The transition from courier to transportation system was already complete under Augustus and so it remained in late antiquity.

This reform is reflected in a provincial edict from Galatia dating from the beginning of Tiberius' reign. The provincial governor Sextus Sotidius Strabo Libuscidianus refers in his edict to directives received from Augustus and Tiberius. This text, first published in 1976, is recorded in both Greek and Latin,[4] and is our only source detailing exactly how the *cursus publicus* worked:

> Sextus Sotidius Strabo Libuscidianus, Legate of Tiberius Caesar Augustus, decrees: 'It is the most unjust thing of all for me to tighten up by my own edict that which the Augusti, one of the greatest of gods, the other the greatest of Emperors, have exactly regulated, that is that no person shall use wagons free of charge.[5]

We can see that the purpose of the edict is to reinforce the stipulation that transportation services had to be paid for:

> Since the indiscipline of certain persons calls for immediate punishment, I have caused to be posted in the various cities and villages a schedule of those services which, according to my decision, must be provided . . . The people of Sagalassos must provide a service of ten wagons and as many mules for the necessary use of the travellers,

> and they shall receive from each user ten asses per schoinum for a wagon and four asses per schoinum for a mule.[6]

The text clearly reveals how Augustus' transportation system worked: it was a burden placed on local populations – in other words, it was a *munus*. The provincial governor designated which services were to be supplied by each city or village. Basically, this meant that the population had to provide transportation services to state officials. In return, they were entitled to compensation on a scale determined by the governor. The system therefore was founded on the principle of forced rental of vehicles and animals, which the local population had to provide at a preferred rate.

Only certain persons were entitled to use the transport thus provided:

> Not all are entitled to use this service, but the imperial Procurator and his son. They may use up to ten wagons or three mules in place of a single wagon . . . ; further, persons travelling on military service may use public transport, both those who have a diploma as well as soldiers stationed in other provinces who are passing through. A Roman senator may use up to ten wagons or, in place of wagons . . . A Roman knight on imperial service shall have three wagons . . . A Centurion may use one.[7]

The edict names explicitly those persons entitled to use the public transportation system: the provincial procurator, senators, *equites*, centurions and generally all persons undertaking imperial or military business, insofar as they have imperial commissions and warrants for the *cursus publicus*. Quarters, however, had to be provided to travellers at no cost: 'Free hospitality must be offered to all members of our household; to persons on imperial business from any province; to the best emperor's freedmen and slaves and their animals.'[8]

Although the payment of compensation for transportation had disappeared by the fourth century at the latest,[9] the system remained much the same for travellers using it. They enjoyed the use of animals and wagons belonging to provincial inhabitants, exchanging them periodically at intervals along the way. In order to make the process more convenient for travellers and less onerous to the people, rest and relay stations were established along the major highways of the empire, in late antiquity called *mansiones* or *mutationes*. Construction of these stations occurred progressively, and the whole programme probably was not complete until the third or fourth century.[10]

The use of state-owned facilities was reserved for those persons specifically authorised by the emperor to do so. This right was documented by writs called *diplomata* or *evectiones*, which detailed what contingent of wagons and animals could be called upon,[11] which route was to be taken,[12] and the period during which the warrant was valid.[13] Only those services noted in the pass could be demanded. It was strictly forbidden to demand any further services or to requisition animals other than at the official stage posts. Furthermore, only the animals belonging to the *cursus publicus* could be used, which sometimes led to delays until fresh animals became available. If travellers were delayed at a way station, they were to wait until fresh animals could be brought up; they were not to demand animals from the nearest peasant. Constantine sanctioned this presumably common abuse with the threat of arrest and indictment.[14] Travellers had to present their warrants when exchanging transport at way stations or to other inspecting officials as required.

The *cursus publicus* was a government transportation system based on obligations placed by the Roman state on private persons. They provided equipment, animals and wagons used by government agents during their travels. In the early empire, compensation had to be paid for this service, but that had fallen into abeyance in late antiquity. The burden of this *munus* then fell completely on the inhabitants, who also had to maintain the way stations and care for the animals.

This system enabled the state to satisfy its most important transportation needs: first, securing state communication; second, transporting government agents; and, third, moving certain goods. The following discussion will show how these three tasks were organised. It will be seen that it was not the personnel of the *cursus publicus* but rather individual government agents travelling on state business that carried out these tasks. The *cursus publicus* contained only those personnel necessary for its administration and for the operation of the way stations. These included veterinarians, wagon-wrights, and grooms responsible for the care, issue and maintenance of the animals at the various way stations.[15] Since neither couriers nor wagon drivers belonged to the *cursus publicus* proper, government agents travelling by means of the *cursus publicus* were not using a transportation system *per se*. They were using an infrastructure based on the facilities of the *cursus publicus* which used animals requisitioned from local individuals or communities.

STATE COMMUNICATION

Who carried the government's messages?

The emperor Caligula was not fated to go down in history as one of its great figures. Concerning his varied efforts in this direction, Suetonius reports the following from the expedition to Britain:

> The only thing that Caligula achieved on this campaign was the taking under his protection of Adminius, the son of the British king Cunobelinus, who came over to him with a small band of followers after being expelled by his father. He sent glowing reports to Rome, however, as though he had subjected the entire island. To this end he urged his *speculatores* to drive their carriage all the way to the forum and *curia* and not to present their dispatches to the Consuls until the Senate was fully assembled in the temple of Mars.[16]

Caligula entrusted the carrying of important dispatches to soldiers, in this case high-ranking *speculatores*. These were probably *speculatores praetorii*, who also acted as imperial bodyguards.[17] Their close proximity to the emperor made them well suited for special tasks, message-carrying among others. Other emperors used soldiers as couriers, including centurions, members of their personal guard[18] and, after the second century AD, the *frumentarii*, although these latter probably only for special missions.[19]

Provincial governors preferred to entrust messages to members of their personal staff such as *beneficiarii*[20] or guard cavalrymen (*equites singulares*).[21] Low-ranking soldiers from the provincial armies were also detached for service as couriers. The morning reports and strength returns of individual units, and other forms of military correspondence, show that in every unit a number of soldiers would be away carrying letters.[22]

Besides soldiers, civilian members of the administration also were occupied with carrying messages. Among these official couriers numbered the traditional message carriers attached to the Roman magistracies, the *geruli* and *viatores*, who were members of the urban decuries of *apparitores* and thus were free citizens.[23]

Probably the largest group of imperial couriers was that of the *tabellarii Augusti* and *cursores*, both professional couriers, as their designations indicate.[24] During the principate, the imperial *tabellarii* and *cursores* were made up of imperial slaves and freedmen. Evidence for them is concentrated at the great centres of administration, in Rome and in provincial capitals throughout the empire, such as Ephesus or Carthage. The *tabellarii Augusti* had been given a military-like organisation by the second century, probably because of their large numbers.[25] Such a structure was common to many large official groups of slaves and freedmen.[26] The *tabellarii* and *cursores* were attached to the various government bureaux – especially within the Roman financial administration, for instance the various financial offices in Rome, the *officium rationum*, the *statio hereditatium*, the *statio patrimonii* or the (*ratio*) *castrensis*.[27] Further, they belonged to the urban Roman administrative departments of the *officium annonae* and the *statio marmorum*.[28] Outside of Rome they could be found on the staff of the imperial procurators.[29] In later times it was particularly the *agentes in rebus*, also known under the titles of *magistrianoi* and *veredarii*, who served as imperial messengers.[30]

Besides soldiers and professional messengers, the Roman state called upon the services of other persons as required. Servants and officials were occasionally commissioned to carry messages. Among these numbered the *servi publici* that, according to Plutarch, the senate dispatched with its decrees after Nero's death.[31] In Egypt, messenger services were performed by *hyperetai*, who were also subaltern officials.[32] Even high-ranking officials from the staff of provincial governors, the so-called *officiales*, were occasionally called upon to perform further unspecified duties as letter-carriers. This was, however, probably an exception to normal practice used only when the governor expected not only delivery of his letters, but also an immediate reply.[33]

The various classes of government couriers just mentioned can be divided roughly into two categories: soldiers and other government personnel belonging to the various administrative offices. They did not belong to the structure of the *cursus publicus*. All of these messengers carried out their tasks for the emperor, for military commanders or for civilian government office-holders. Government couriers thus formed a part of the Roman administration.

From Pliny's correspondence with Trajan we learn that the issue of warrants for the use of the *cursus publicus* was extremely restricted and that such warrants were only issued to messengers or other officials on a case-by-case basis and for important matters alone. For example, the governor of Bithynia writes to the emperor informing him that he had issued such a warrant to a messenger of King Sauromates in order to accelerate his journey to Trajan: '*ex causa festinationem tabellarii . . . diplomate adiuvi*'.[34] It follows that even official state couriers did not always travel via the *cursus publicus*. If their mission was not particularly urgent or not of the highest priority, they had to travel on foot or use vehicles not exchangeable at the way stations.

Two special cases seem at first glance to contradict the statements just defined. A series of late Roman papyri from Egypt seems to refer to letter-carriers belonging to the *cursus velox*, a section of the *cursus publicus* established in the fourth century for express deliveries.[35] The exact title of such an official was *haliadites etoi grammatephoros* or 'boatman or letter-carrier' of the *cursus velox*: we might infer from this functionary's title that he did in fact belong to the staff of the *cursus publicus*.[36] Since no comparable official title is known from the rest of the empire, this might be a peculiar institution within the Egyptian transportation system. But in four papyri from a period of more than sixty years we find exactly the same title, where the function as sailor is always named first.[37] Further, the earliest example of this post from the year 300 gives the description as 'sailors in the service of the *cursus publicus*'.[38] It follows that the main function of these liturgical officials was that of boatman, and they would have plied their trade in shallow-draught Nile boats as part of the *cursus velox*. This system would have been particularly appropriate for Egypt, where the centres of population lay close by the banks of the Nile. The fact that these boats operated along one main route made it sensible for them to transport not only the official travellers who were allowed to use the *cursus publicus*, but internal administrative mail of the province of Egypt as well. The requirement for reliable and frequent delivery of official correspondence was particularly important in Egypt, with its traditionally complicated bureaucracy.[39] A similar water-transport service for official travellers was provided by the state in the Adriatic Sea. Tacitus shows that *biremes* of the Roman imperial navy were used to transport officials between Italy and Dyrrachium.[40]

A further deviation from the communications system already outlined deals with the delivery of messages between military installations, particularly on the military frontiers. A true courier system existed in the military zone, so-called *equites dispositi* being located at designated stations. These mounted messengers delivered their dispatches in a system of relays, thus maintaining high-speed communications between headquarters, scattered detachments and field forces.[41] This, however, was a purely military institution, having nothing to do with the *cursus publicus*, so the two should not be confused. The military courier service fulfilled strictly military requirements and it did not exist outside the military zones, and indeed was not universal there. For supra-regional communications involving, for instance, correspondence with neighbouring armies, the governor or the emperor detached soldiers who could then use the *cursus publicus* if the urgency of the mission warranted it.

TRAVEL BY GOVERNMENT OFFICIALS

Besides state couriers, other government officials were entitled to use transportation provided by the *cursus publicus* when in the possession of a valid warrant. That the travelling officials themselves and not the members of the *cursus publicus* were the users of animals and wagons is shown clearly by a statute of the emperor Constantine. In 316, he forbids the use of cudgels to drive the beasts, whips only being permissible.[42] As usual with imperial laws there is a punitive clause providing for punishments in case of abuse. In the case of this law, the threat of punishment is directed against senior and low-ranking government officials, a *promotus* and a *munifex*. It follows that officials either drove the vehicles themselves or were held personally accountable for their use of *cursus publicus* transportation facilities, meaning that they had personnel in their retinues for this purpose.

Various sources indicate the presence of such drivers on the staff of travelling government officials. The earliest such document is a second-century inscription from Carnuntum containing the following text: 'Sacred to Epona Augusta, those in charge of looking after the baggage animals and mules of Claudius Maximus, the imperial propraetorian legate, gladly, willingly, and deservedly fulfilled the vow'.[43] Claudius Maximus is known as provincial governor of Pannonia superior between 150 and 154.[44] The first editors of this inscription, Jobst and Weber, saw the *superiumentari et muliones* as officials of the *cursus publicus*.[45] Werner Eck disagreed, with good reason, showing that these *superiumentari et muliones* would never have called themselves the *superiumentarii et muliones* of the governor Claudius Maximus if they had worked for the *cursus publicus*. They must therefore be regarded as members of the governor's staff.

A similar document from AD 217 exists from Apulum, the capital of the province of the Tres Daciae. The inscription was dedicated by a *superiumentarius*, once again, to the goddess Epona and to the well-being of the provincial governor.[46] The servant designates himself as *superiumentarius eius* (of the governor) and this shows that he was indeed a member of the governor's staff.

Such personnel were common in late antiquity as well. The Christian writer Eusebius preserves a letter from Constantine to Bishop Chrestus of Syracuse, inviting him to a synod at Arles. In it, Constantine authorises him to use the *cursus publicus* for his journey and to take two other churchmen and three servants along, such as might be serviceable to him along the way.[47] A law from the Theodosian Code concerning the *cursus publicus* shows that accompanying personnel were necessary, not just for protection, but because of the hardships of travel.[48]

TRANSPORT OF GOODS

Freight played a relatively unimportant role in the *cursus publicus*. For the principate, the Galatian edict indicates that this freight consisted primarily of baggage belonging to government officials. Freight transport on a larger scale, such as supplies for the army, does not seem to have been organised using transport provided by the *cursus publicus*, because the facilities were not sufficient for such provision.[49] The same seems to have been true in late antiquity. The *cursus publicus* was not used to transport taxes in kind.[50]

During the late imperial period, the *cursus publicus* was used to transport tax money and a number of products from state industry, such as weapons or clothing for the army[51] or the imperial court.[52] Imperial laws show that only the officials of the financial agency responsible organised the transportation of these goods: the officials of the *comes sacrarum largitionum* and the *comes rerum privatarum*, the chiefs of the two imperial treasuries.[53] They received warrants for the use of horses and wagons of the *cursus publicus* whenever they required them.[54] Other officials did not enjoy this special dispensation. These generally received an annual restricted allotment of *cursus* warrants. For example, according to a law of AD 316, provincial governors were entitled to only two *evectiones* annually. The emperors declared themselves willing, however, to issue additional warrants to provincial governors in urgent cases or for reports being submitted to the court.[55]

One unusual consignment carried for the emperor Constantine illustrates once again that it was not the personnel of the *cursus publicus* who carried out freight transport, but rather the responsible official himself. In a letter to Eusebius, Constantine asks that he send fifty of the most useful theological books to him at Constantinople, and allows him two government wagons for the purpose. A deacon of Eusebius' church was to demand these, presenting Constantine's letter to Eusebius as authorisation.[56]

CONCLUSION

The purpose of the *cursus publicus* was to make possible official government travel. This was accomplished by organising transportation at particular places along certain routes, which then were used in relays. The conduct of business was left to the system's users. These were, on the one hand, state couriers belonging to government bureaux or the military, and, on the other, government officials travelling on business throughout the empire. Finally, certain goods were transported through the endeavours of state officials. The *cursus publicus*, therefore, was not a delivery service like a post office, but the infrastructure which allowed for the provision of official transportation.

NOTES

1 Hudemann (1878); Holmberg (1933); Hirschfeld (1963); Pflaum (1940); Stoffel (1994); Kolb (2000) 49–226.
2 See Eck (1979) 88; Stoffel (1994) 3.
3 Suet., *Aug.* 49.3; cf. Hirschfeld (1963) 190–204; Holmberg (1933) 37–42; Pflaum (1940) 210–45; Eck (1979) 88–9.
4 Mitchell (1976) 106–31 = *SEG* XXVI 1392 = *AE* 1976, 653; for the dating not after AD 19, see Mitchell (1986) 26–33. The following quotes rely on the Latin text of the inscription.
5 *SEG* XXVI 1392, 1–4: '*Sex(tus) Sotidius Strabo Libuscidianus, leg(atus) Ti(beri) Caesaris Augusti pro pr(aetore) dic(it). Est quidem omnium inquissimum me edicto meo adstringere id quod Augusti alter deorum alter principum maximus diligentissime caverunt, ne quis gratuitis vehiculis utatur*'.
6 *SEG* XXVI 1392, 4ff. (with ZPE45 1982, 100): '*Sed quoniam licentia quorundam praesentem vindictam desiderat, formulam eorum quae* [*pra*]*estari iudicio oportere in singulis civitatibus et vicis proposui servaturus eam aut si neglecta erit vindicaturus non mea tantum potestate sed principis optimi a quo id – in ip*[*sum*] *mandatis accepi maiestate . . . Sagalassenoss*[*o*] *ministerium carrorum decem et mulorum totidem praestare debent ad usus necessarios transeuntium, et accipere in singula carra et in singulos schoenos ab iis, qui utentur, aeris denos, in mulos autem singulos et schoenos singulos aeris quaternos.*'
7 *SEG* XXVI 1392, 13ff.: '*Neque tamen omnibus huius rei ius erit, sed procuratori principis optimi filioque eius, usu da*[*to us*]*que ad carra decem aut pro singulis carris mulorum trium . . . praeterea militantibus et iis, qui diplomum habebunt, et iis, qui ex alis provincis militantes commeabunt ita ut senatori populi Romani non plus quam decem carra . . . equiti Romano cuius officio princeps optimus utitur ter carra . . . centurioni carrum.*'
8 *SEG* XXVI 1392, 23: '*Mansionem omnibus qui erunt ex comitatu nostro et militantibus ex omnibus provincis et principis optimi libertis et servis et iumentis eorum gratuitam praestari oportet.*'
9 Other documents show this procedure only until the third century AD: *P. Lond.* III 1171 v (AD 42); *OGIS* 665 (AD 48); *IGL*Syr. V 1998 (AD 81–96); *SEG* XXXVII 1186 (AD 211–14).
10 For example, *CIL* III 6123 = *ILS* 231add.; *CIL* X 7200 = *ILS* 5905.
11 *CTh* 6.29.2.2 (AD 357): '*ne quis citra evectionis auctoritatem moveat cursum vel amplius postulet, quam concessit evectio*'; 8.5.14 (AD 362); 22 (AD 365); 24 (AD 365); 27 (AD 365); 29 (AD 368[367]); 35 (AD 378); *Nov. Iust.* 30.7.3.
12 *CTh* 8.5.25 (AD 365).
13 *CTh* 7.12.2 (AD 379); *CTh* 8.5.9 (AD 357); 12 (AD 362); 27 (AD 365).
14 *CTh* 8.5.1 (AD 315).

15 *CTh* 8.5.31 (AD 370).

16 Suet., *Caligula* 44.2: '*Nihil autem amplius quam Adminio Cynobellini Britannorum regis filio, qui pulsus a patre cum exigua manu transfugerat, in deditionem recepto, quasi universa tradita insula, magnificas Romam litteras misit, monitis speculatoribus, ut vehiculo ad forum usque et curiam pertenderent nec nisi in aede Martis ac frequente senatu consulibus traderent.*'

17 Clauss (1973) 46–53.

18 Tacitus, *Hist.* I 67; II 98; IV 37; Herod. VII 6. 5; *Tab. Vindol.* II 263; Tacitus, *Hist.* II 73 shows *speculatores* as couriers of the emperor Vitellius.

19 See Kolb (2000) 290–4. For the contrary view see Rankov (1990) 180: 'The most fundamental role of the *frumentarii* seems not to have been espionage, but the carrying of messages between the emperor and the provinces'; similarly Eck (1995) 69.

20 SHA, Hadrian 2. 6.

21 *P. Oxy.* VII 1022 = Fink, *RMR* nr. 87, 26 (AD 103).

22 For example, *P. Dur.* 82 = Fink, *RMR* nr. 47 col. II 1–7: '*iiii kal(endas) april(es) n(umerus) p(urus) mil(itum) cal(igatorum) dccc{c——} / coh(ortis) XX Palm(yrenorum) Severia{na}e A{l}e{xandrianae} / Iulius Rufinianus tribunus ..vessus ..{——} / missi ad hordeum comparandum m{il(ites) . . . } in h(is) eq . . . {——} / missi in prosec(utione) hordiator(um) mil(ites) {.} i (centuria)Mariani .{——} / reversi q(ondam) d(e)p(putati) Adatha mil(ites) ii (centuria) Nigrini Iul(ius) Zabdibolus{——} / reversi q(ondam) d(e)p(putati) ad praet(orium) praesidis cum epistul{i}s . . . {}*'.

23 For *geruli*, see Mommsen (1878) 366; for *viatores*, ibid. 360–2; Habicht, *RE* VIII A2, 1958, 1928–40; Kunkel and Wittmann (1995) 123–5; cf. Waltzing (1895) 416; Purcell (1983) 125–73.

24 *CIL* V 6964 = ILS 1701 (Turin): '*Ex tabellar(iorum) Aug(usti) stat(ionis) Taur(icae) l. m. d.*'; *CIL* VIII 1878 = *ILS* 1709; *CIL* X 1741; *IGRR* IV 1221; *IG* V 2, 525. For *cursores*, see for example, *CIL* VI 241. 8800. 9317; *AE* 1899, 103; Hirschfeld (1963) 200.

25 The inscriptions give the following ranks of *tabellarii*: *praepositi*, *decuriones*, *optiones* and *tesserarii* in *CIL* VI 410. 746. 8424. 8445. 8505. 9915. 37,766; *IvEph* 696A. 2200. 2222B. 4112; *SEG* VI 594.

26 For example, *praepositus unctorum CIL* VI 8582; *praepositus lecticariorum CIL* VI 8874; *decurio lecticariorum CIL* VI 8875; *decurio unctorum CIL* VI 9093; see further *optiones* with the *officinatores monetae*, cf. Weaver (1972) 228.

27 *CIL* VI 8424a. 8445. 8505. 8526; cf. Hirschfeld (1963) 201–2; Pflaum (1940) 318–26.

28 *CIL* VI 410. 8473.

29 *IvEph* 855: '[*adiutores*] *et tabel(larii) q(ui) s(unt) sub cura* [. . .] *li Firmi proc(uratoris) Aug(usti)*'; also 2200A. 4112; *CIL* VIII 12623–35. 12906. 12908–11.

30 Jones (1964) 578–80; Hirschfeld (1913) 576–612; Giardina (1977); Clauss (1980) 23–51, especially 45–51; Palme (1994) 43–68.

31 Plut., *Galba* 8.4.

32 For their task as letter-carriers see *BGU* I 226 (AD 99); II 578 (AD 189); cf. Strassi (1997) *passim*, especially 47–8.

33 *P. Lips.* inv. 366 = *SB* XVIII 13251; cf. Kramer (1986) 33–9.

34 Pliny the Younger, *Epistulae* 10.64.

35 Cf. Thomas (2000) 194–5.

36 *P. Flor.* 39 (AD 396). The term derives from the word *haliades* which describes a small boat used by officials (military and civilians) and private persons; for sources and references cf. *CPR* X 2.
37 *P. Oxy.* LI 3623 (AD 359); *PSI* X 1108 (AD 381); *P. Flor.* 39 (AD 396); *P. Oxy.* LV 3796 (AD 412).
38 *P. Panop. Beatty* II, 275 (AD 300).
39 With every village submitting monthly reports to the financial administration, a periodic system of official mail was vital.
40 Tacitus, *Ann.* IV 27.1; Starr (1993) 23 and 178; Eck (1997–8) 339–46.
41 In this way Davies (1989) 59 explains the fifteen *dispositi* in the province of Syria Coele: *P. Dura* 100 (AD 219) and 101 (AD 222)= Fink, *RMR* 1. 2; cf. Fink (1971) passim, s.v. 'dispositus'. Caesar and Pompeius, though, had used such mounted soldiers during the civil war: Caesar, *Bell. Civ.* 3.101: '*nuntii de Caesaris victoria per dispositos equites allati*'; Caesar, *Bell. Hisp.* 2: '*tabellarios, qui a Cn. Pompeio dispositi onmibus locis essent, quo certiorem Cn. Pompeium de Caesaris adventu facerent*'.
42 *CTh* 8.5.2 (AD 316): '*Quoniam plerique nodosis et validissimis fustibus inter ipsa currendi primordia animalia publica cogunt quidquid virium habent absumere, placet, ut omnino nullus in agitando fuste utatur, sed aut virga aut certe flagro, cuius in cuspide infixus brevis aculeus pigrescentes artus innocuo titillo poterit admonere, non ut exigat tantum, quantum vires valere non possunt. Qui contra hanc fecerit sanctionem promotus, regradationis humilitate plectetur: munifex poenam deportationis excipiat*'.
43 Eck (1992) 207–10: '*Eponae Aug(ustae) sac(rum) Cl(audii) Maximi leg(ati) Aug(usti) pr(o) pr(aetore) superiumentari et muliones v(otum) s(olverunt) l(aeti) l(ibentes) m(erito)*'.
44 *CIL* XVI 99.104; *PIR*[2] C 933; Thomasson (1984) 157 nr. 56.
45 Jobst and Weber (1989) 349–58 = *AE* 1991, 1308.
46 Stein (1944) 65 = *IDR* III 5.71: '*Epone sancte pro salute C(ai) Iuli Septimi Castini, leg(ati) Aug(usti) pr(o) pr(aetore) III Daciar(um), Libella superi{u}mentarius eius [votum s]olvit.*'
47 Eusebius, *HE* 10.5.23.
48 *CTh* 8.5.4 (AD 326): '*ad tutelam vitae vel laborem adeundum itineris*'.
49 Cf. Herz (1988) 60.
50 See Kolb (2000) 227–47.
51 *CI* 12.50.22 (AD 467/8): *arma*; *CTh* 8.5.33 (AD 374): *vestes militares*.
52 *CTh* 8.5.48 (AD 386): *lineae vel amictoriae*.
53 *CTh* 8.5.13 (AD 362); 18 (AD 364); 40 (AD 382); cf. *CTh* 8.5.20 (AD 364).
54 *Not. Dig.*, Or. 13.35; *Not. Dig.*, Or. 14.15: '*comes rerum privatarum quotiens usus exegerit*'. But see Delmaire (1989) 249, who misunderstood these sources and thought that these officials had the right to give out *evectiones* by themselves.
55 *CTh* 8.5.12 (AD 362): '*Sed ut necessitates publicae impleantur, vicariis denas vel duodenas evectiones manu mea perscribtas ipse permittam, praesidibus vero binas annuas faciat vestra sublimitas, quibus ad seperatas provinciarum secretasque partes necessariis ex causis officiales suos dirigere possint. Sed his quoque nostra etiam mansuetudo evectiones singulas dabit, ut ad nos referre possint, cum fieri necessitas quaedam exegerit*'.
56 Eusebius, *VC* 4.36.4.

6

TRANSPORT AND TRAVEL ON THE COLUMN OF TRAJAN

Jon Coulston

INTRODUCTION

The sculptures of Trajan's Column in Rome have been studied in many ways and employed in a great diversity of roles over the centuries, from acting as an inspiration for Renaissance artists, to serving as a source of information for forest species in ancient Dacia.[1] Traditionally, for historians, the literalist approach of treating the helical frieze as a straight translation of Trajan's lost *commentarii* into stone has been almost irresistible.[2] However, in most quarters this has given way to more subtle appreciations, concerned less with 'narrative' than with questions of imperial propaganda and socio-political presentation.[3] For Roman archaeologists the temptation has been to 'dip into' the reliefs to illustrate discussions of military practices. Here, too, more discernment is now displayed and the column has been retired from primary-source status in the face of new artefactual and structural evidence.[4]

Recording and study of Trajan's Column from scaffolding in the 1980s has allowed scholars to concentrate on the architectural design and construction of the monument, and on the practicalities of sculptural planning and execution.[5] This work has elucidated the generation and composition of the reliefs, allowing their content to be examined and evaluated in their proper context.

Predictably, carving was executed *in situ* on the assembled column and it proceeded from bottom to top, but the helical frieze was not laid out ahead of work, as is indicated by the irregular course of the dividing band. This latter is in complete contrast to the later practice on the Marcus Column, which had Trajan's Column as a model. Crucially for the purpose of modern study, a secure internal chronology of work-order is imposed on the sculpting project, starting at the bottom. The frieze is divisible by figural and scenery composition into individual 'scenes', as defined by Cichorius.[6] Some larger exceptions aside (e.g. IV–V, XXIII–XXIV, LXXV), most of these may represent a work-unit encompassable by one sculptor. Many scenes are formulaic genre compositions without any specific 'historical' reference (march, construction, battle, *adlocutio*, etc.).[7]

However, some scenes were linked by content and propaganda message in vertical axes up the column shaft cutting across the helical progression of the frieze in order to aid the viewer below to grasp the essential propaganda messages.[8] This is especially clear with the bridge-crossing scenes discussed below (IV, XLVIII, CI, CXXXI). Human figures were defined early according to 'figure types' in order to simplify the depiction of large categories.[9] As work progressed sculptors made inevitable mistakes and confusions between types, but these diminished with experience. Some minority types evolved during work and are never quite the same each time they occur (notably Roman archers).

Much sculptural detail is distributed in such a way as to indicate work executed by sculptors carving alongside each other and moving vertically up the shaft. Details of military equipment, for example specific shield-blazons, cuirass-fittings or helmet-types, cluster in groups of scenes, often vertically across two or three windings of the helix. This indicates that composition was done at the stone face and refutes the traditional assumption that sculptors followed detailed 'cartoons'.[10] Rather, it was an evolving process generated by the experience and penchants of individual sculptors. This elevated the role of such men from mere copyists working under a putative *maestro* to composers taking a full part in relief generation.[11]

The overall column project would thus have progressed from architectural design and execution to a consultation of the *commentarii*. The latter probably dictated the narrative structure of two wars divided into seasonal campaign periods. Once the progress of 'events' was determined, the helix was perhaps laid out on a model with important scenes labelled and vertical axes scenes marked out.[12] Main figure types were then formulated. Work on carving the helix proceeded from bottom to top with composition of figures, scenery and details evolving during the process. Lastly, paint and gilding were applied.[13]

This chapter employs these new perspectives to study the depictions of travel and transport in the helical frieze of Trajan's Column on two main levels. On the one hand it will examine what the column actually represents set within the context of other evidence for Roman travel, transport and transport installations. Thus such matters as marching soldiers, pack-animals, tents, wheeled vehicles, mounts, shipping, bridges and ports will be discussed. On the other hand, it will explore the propaganda and cultural meanings of the travel and transport imagery presented to the viewer by the sculptors. Travel is inherent to the frieze which depicts a series of journeys made by Trajan and his armies (not always in company) to bring force to bear against Decebalus and his Dacians, to conquer Dacia, and to garner the treasure which helped finance the wondrous building complex within which the column stood.[14] It will be clear that various forms of transport were crucial to the campaign movements of Roman armies. However, these were primarily depicted on the column not so much as part of a realistic narrative as to help define the nature of Trajan's victories which were won through his army's inherently Roman culture and capabilities.

TRAVEL AND TRANSPORT ON LAND

On the column frieze Roman armies advance repeatedly into *barbaricum* and they are represented as columns of infantry, mounted and dismounted cavalry, pack-animals and wheeled transport.[15] The advance at the beginning of the First Dacian War is the most impressive representation in the 'marching column' genre, but at this early stage the first appearance of the citizen soldier figure-type displays some features which were either confused or not thereafter used (e.g. multiple waist-belts, very ornate standards). It is also the only time that soldiers are shown marching with full equipment (Figure 6.1). Bags, wineskins, *paterae* and other vessels are slung on the end of a pole carried over each man's shoulder. Only one such 'pack' of belongings appears elsewhere, and that on a ship (XXXIV). Experimentation has demonstrated that this pole-end carriage is quite impractical because the centre of gravity would be too high up and the pole would be uncomfortable to hold. More likely the items would have hung directly over the shoulder, perhaps conveniently slung from the soldier's *dolabra*.[16]

Why then were these 'packs' depicted in this way? The answer is perhaps partly that the sculptor wished to depict clearly an army setting out with all its equipment at the beginning of the war, but in an elaborate fashion which was not adopted for later such contexts (XLVIII, CI). Most significantly, this scene comes at the point where the frieze having wound once around the shaft

Figure 6.1 Trajan's Column, scenes III–V: river god, bridge and marching column with 'packs'. Photograph J.C.N. Coulston

now mounts its own 'tail' to start the helical progression. Going from horizontal to diagonal involved the creating of a long triangular space above the soldiers' heads, which the 'packs' were conveniently, but unrealistically, created to fill. Further on (V) this same function is served by the tall, elaborate praetorian *signa*, and by the long-shafted spears of dismounted cavalry which are also very rare on the frieze. In the same scene *cornua* have long, peltate-terminal crossbars projecting up into the overhead space, a realistic feature, but one which is also not seen elsewhere on the column.

In a number of scenes citizen troops march bareheaded with their helmets slung down over their chests.[17] Each helmet would have had its cheek-pieces folded up and tied inside the bowl, and the whole would have been suspended from the carrying handle which is so commonly found on the neck-guards of principate helmets.[18] This method of carriage is accurately depicted on Trajan's Column to judge from other monuments depicting troops at rest or on the march.[19]

The first example of a pack-mule on Trajan's Column appears in scene XV (Figure 6.2). The beast is crammed into a small space. It has shoulder- and breacher-straps and a pack-saddle with rectangular side-frames, but no visible load. The man accompanying this animal is strangely stripped to the waist in a section of the helix where all the soldiers are fully clothed and cuirassed. Scene XLIX figures two pack-mules carrying trussed bundles on supporting side-frames. A pack-mule with shoulder- and breacher(?)-straps appears in each

Figure 6.2 Trajan's Column, scene XV: forest clearance, small bridge, and pack-mule with attendant (top right). Photograph J.C.N. Coulston

of scenes CVI and CVII. The first bears a helmet and a pair of oval shields slung on either side; the second is being loaded or unloaded with a pair of trussed bundles. In the much-reproduced scene of soldiers reaping corn (CX) three mules stand barebacked with only head-harness and reins, perhaps waiting to be burdened with sacks or baskets of corn. Towards the end of the second war when Decebalus' treasure has been captured three horses or mules stand with attendant soldiers (CXXXVIII). On two, fringed saddle-blankets are visible, and on all three paniers are shown full of *paterae* and other presumably precious metal vessels.[20]

Pack-mules are not very frequently depicted in Roman art, perhaps because they were so ubiquitous and thus too commonplace a subject. They appear in landscape painting and mosaics, and as terracotta figurines. None seem to be present on the Marcus Column, but they do figure on the Arch of Constantine and the Column of Arcadius.[21] Mules were an extremely efficient transport animal, highly valued in Roman Italy and widely employed by ancient armies.[22]

The trussed bundles have generally been identified by scholars as packed-up tents, the *papilio* of Roman military word-play.[23] They may also represent wrapped bundles of equipment or supplies, but whatever they are, they are substantially scaled down in size. Similar items are occasionally manhandled on or off ships or into camp.[24] Pitched tents frequently appear in the camps on the frieze, larger ones presumably for officers, smaller for *contubernia* (Figure 6.3).[25] The Roman army lived 'under leather' not canvas,[26] and enough different portions of leather tents have survived in wet archaeological contexts to allow reliable reconstruction of the eight-man form.[27] It was not the low-walled 'hike-tent' of earlier reconstructions based on misinterpretation of the column,[28] but the higher-walled 'Vindolanda' type in which people could stand up (Figure 6.4). Careful examination of tents on the frieze with allowance for masking by camp defences reveals that this is precisely what the column depicts. A tent weighed *c.* 18–20 kg when dry and required some seventy goatskins for its manufacture.[29] It is a sobering thought that at least 46,000 goats would have died to put one legion *sub pellibus*.[30]

Tents were not an absolute necessity for marching troops and not all campaigning armies in history carried them, but they did have three great advantages, especially for pre-modern forces. First, they helped the troops to rest under cover, reducing cumulative exhaustion, particularly in bad weather and cold campaigning. Second, they helped to preserve vulnerable equipment in good order by protecting it from the elements. They also provided a place for drying out items which were particularly susceptible to rain or wetting during the crossing of water-courses: that is, any equipment using wood and/or leather, notably shields, composite bows and artillery torsion springs.[31] Third, the pitching of large numbers of tents in a sizeable 'marching camp' necessitated, and was inherent to, the camping discipline practised by the Roman armies from an early date. This is clear from the literary record and from the layout of the barrack-blocks in more permanent military installations which had evolved

Figure 6.3 Trajan's Column, detail of scene CXXVIII: tents within a military camp. Photograph J.C.N. Coulston

Figure 6.4 Leather tent reconstructed by the Ermine Street Guard. Photograph M.C. Bishop

from surveyed tent-lines.[32] After the late Roman period such discipline was not developed again in Western warfare until the seventeenth century.[33] The Roman tent never stood alone, as it were, but represented both hard campaigning and the civilised, disciplined methods of Roman warfare. It goes almost without saying that no Dacian tents are depicted on Trajan's Column.

One feature of Danubian warfare which might confidently be expected is missing from the reliefs of Trajan's Column. There are no pack-camels. Dromedaries were commonly employed by the army throughout the eastern Roman provinces as beasts of burden and, less commonly, as cavalry mounts.[34] They were very efficient carriers, bearing twice as much as a mule and up to 70 per cent more than the load of a pack-horse, at comparatively greater speed.[35] For these reasons the Ottomans made great use of camels in the Danubian theatre, both Bactrian and dromedary, notably during their thrusts up to Vienna in 1529 and 1683.[36] Camels even made it as far north-westwards as Frankfurt (Germany) in diplomatic entourages, causing a *frisson* of interest in the west.[37] As a long-term result of Turkish use they are still to be found in the Balkans today.

Whenever Roman military formations travelled westwards into Europe they would have brought their camels with them.[38] For example, *legio III Gallica* was transferred from Syria to Moesia late in Nero's reign, and the civil war saw various other Syrian formations on the Danube.[39] Eastern troops certainly participated in Trajan's Dacian wars, including legionary and auxiliary *vexillationes*, auxiliary regiments of *sagittarii*, and perhaps irregular archer formations.[40] Camels were allowed space in Pseudo-Hyginus' design for a Danubian field-army camp, and camel-bones have been found on Roman military sites as far west as Vindonissa in Switzerland.[41] Bactrian camels were also known in the Roman world.[42]

All the vehicles represented in Roman use on Trajan's Column are two-wheeled carts with a pair of draft-animals (Figure 6.5). Despite the many early marches, they do not appear until scene XL, where two mule-drawn box-carts bear artillery. A little further on the bridge-march, scene XLIX shows one mule- and one ox-drawn cart, both loaded with military equipment. In scene LXII, a landscape of mountains and buildings, there are three carts, one with mules, two with oxen. One of the latter is damaged but the other bears three barrels supported on cradles. The mule-cart has two barrels. There is one cart with mules and an artillery piece in scene LXVI and one mule-cart bearing military equipment in LXXXVIII. In the great march scenes of the second war, mule-drawn carts appear in the company of pack animals in two adjoining scenes (CVI–CVII), the first carrying shields, the second bearing three trussed bundles which are being loaded or unloaded in camp. Lastly, two unhitched carts stand in a camp in scene CXXIX, each loaded with two barrels. All the carts have spoked rather than solid wheels, but numbers of spokes are rather random.[43]

Light, two-wheeled carts drawn by two animals were employed throughout the Roman provinces and the type of draft-animal varied with function: horses

Figure 6.5 Trajan's Column, detail of scene LXVI: artillery-piece mounted on a mule-cart. Photograph J.C.N. Coulston

for speed, mules for endurance and oxen for heavy traction. The latter were slow, perhaps the slowest element of a marching army column, but they were strong and easiest to feed with locally available fodder. Modern estimates of capacities vary, with a two-mule cart load of 250–300 kg and a two-oxen cart of 380–400 kg. Given an individual pack-mule capacity of 100–35 kg, even with the most conservative figures there were load advantages with wheeled vehicles, but all was dependent on the type of terrain and quality of roads and bridges involved.[44]

Two-wheeled carts drawn by mules or oxen appear on the triumphal friezes of the Arch of Trajan at Benevento and the Arch of Severus in the Forum Romanum, and on the Column of Marcus Aurelius.[45] Mule or ox-carts are seen elsewhere in Roman iconography transporting large wine-barrels and, sometimes, massive wineskins.[46] Apart from loads, there is nothing inherently military about the vehicles and draught-animals on Trajan's Column. The sculptors could have modelled them on what they saw in Rome, even on what they observed delivering materials to the Forum Traiani construction-site.

Rhenish and Danubian funerary monuments represent four-wheeled wagons in Roman military use.[47] However, only once do these heavier vehicles appear on Trajan's Column, and then they are in barbarian, not Roman use (XXXVIII). Three unhitched box-wagons carry various items including a *cantherus*, a wineskin and a barbarian *draco* standard. The latter make their owners' identity clear, and the vehicles have been related by commentators to the Danubian barbarian wagons mentioned in the ancient literary sources and to the four-wheeled conveyances shown on the Tropaeum Traiani metopes at Adamklissi in Romania.[48] It is unclear whether this scene represents barbarian peoples migrating with wagons, as do the Adamklissi sculptures, or an incursion into Roman territory by raiders using wagons to transport booty (represented by the *cantherus*).

The artillery pieces on the column all belong to one type characterised by a pair of narrow, conical-topped drums linked by an arched horizontal strut, and flanking a projecting 'beam', all set on a stand (see Figure 6.5 above). All these elements correspond with surviving technical texts, particularly the late Roman *Cheiroballistra* of Heron. The 'beam' is the 'slider' (*diostra*) used to pull back the string of this torsion crossbow. Richmond doubted the realism of the arched strut (*kamarion*), but it has been vindicated by an example from Orşova in Romania, which may also be compared with treatise illustrations.[49] On the other hand, misunderstood and inconsistent details were also incorporated. Sometimes the *kamarion* is too schematically rendered as a segment, rather than as a semicircular arch in a horizontal strut (LXVI). The stand has a reasonable elevation for soldiers to operate the piece where it is in the open (LXVI) and it is not depicted in some cases where it would be masked by a wall or the side of a cart (XL, LXVI). However, stands are also shown as low, squat triangles on other walls or carts solely in order that they may still be displayed (LXVI). In most cases the arms of the bow are entirely omitted or, in one instance, a thick horizontal member is extended to either side of the drums (LXV). Arrows but no sliders are shown projecting from the fronts of machines in scene XL.

Each appearance of artillery-pieces on the frieze differs from the rest in detail and proportions. None could have functioned to shoot projectiles if the detail provided is taken literally. Therefore, the sculptors were applying empirically observed elements without any real understanding of their practical functions. The same confusing combination may be observed in the siege(?) machine of scene CXIV.[50]

The basis of the frieze artillery was a machine with sinew torsion coils housed in copper-alloy drums to protect them from moisture damage. A metallic frame was proportionally wider and lower than earlier Roman and Hellenistic wooden models and allowed powerful, yet small and light machines to be constructed. The *kamarion* arch was used in aiming the weapon, particularly in elevation, while the absence of a large, enclosing frontal panel with a small opening allowed short quarrels rather than long arrows to be loosed.[51] Modern commentators consider this type of artillery piece to have been an innovation of the Domitianic–Trajanic period precisely because the earliest evidence for it is provided by Trajan's Column. This may indeed be true, but some caution must be exercised in such cases of prior source 'silence'. The '*lorica segmentata*' was in the past similarly thought to have been a Trajanic invention based on its depiction on the frieze, whereas other evidence may now be adduced for pushing it back into the Augustan period.[52]

Artillery pieces usually stand with citizen troops in reserve, not directly participating in battle. In contrast, the cart-mounted artillery in XL is more active, as the arrows show, demonstrating that it was being shot from the carts, not merely being transported on them. This scene is in one of the few battles which involve citizen troops fighting. Artillery is excluded from march-scenes but, contrary to expectation, it is also absent from sieges. It is the preserve of citizen troops, thus tied to them in context. This seems to reflect the reality of provision in the Roman army with artillery integral to praetorian and legionary organisation, but normally absent in the *auxilia*.[53]

The column's weapons were long-range, point-accuracy pieces which would have been used for specific, individual targets, such as barbarian leaders or individuals attempting to threaten Roman besiegers from a wall-top.[54] Positioned on the flanks in battle they would have supported a legionary line, and mounted on carts (*carroballistae*) they could have kept up with advancing Roman infantry. Accurately directed bolts with incendiary heads would have been particularly effective in sieges.[55] Light and mobile cart-mounted artillery continued in use until at least the sixth century.[56] No larger, stone-throwing catapults are figured on the column. Perhaps they were not used in Dacia, and barbarian hilltop fortifications were assaulted without artillery wall-breaching. Alternatively, the sculptors were unfamiliar with them in Rome. The artillery used by *praetoriani* and *vigiles* in the capital may have provided general models and the specific, accurate details.[57]

Once, Dacians are shown serving an artillery piece (LXVI) and this might be taken to corroborate the use of captured Roman artillery by the barbarians.[58] However, the weapon is pointing in diametrically the wrong direction, up the helix towards the viewer's right, away from the advancing Romans and aimed at Dacian forces. This excites the suspicion that what is involved here is not a contact with the literary sources but a sculptor's mistake. Perhaps he confusedly mixed up the attributes and functions of two figure types, as clearly happened elsewhere.[59]

Carts and artillery pieces are closely associated on the frieze and they cluster in a group of scenes on the ninth and tenth spirals (counting up the south-east face). Taken together, the thirteen pack-animals and thirteen carts display very little uniformity of load and construction from scene to scene. This is also a compositional feature observed of certain figure-types and equipment details, which evolved as work progressed up the column shaft. While some features are corroborated by other sources, it seems that visual variety was a significant factor in the sculptural process.

The remaining mode of land transport on the column to be examined is mounted travel. A mule is ridden in scene IX by the first barbarian to appear on the frieze.[60] Roman cavalry horses are led or held on occasion; more commonly they are depicted ridden in transit or in battle (Figure 6.6).[61] Clearly the horses themselves have been scaled down in size, a practice displayed also on figural cavalry gravestones where the mount had to be fitted into a narrow *aedicula* or

Figure 6.6 Trajan's Column, detail of scene CIV: horse with 'triplet' straps hanging from saddle-horn. Photograph J.C.N. Coulston

field.[62] On the column the natural bodily bulk of horses was not allowed to dominate or obscure their owners. Thus these animals cannot be employed in discussions about the size of Roman breeds, however small they were by the standard of modern cavalry mounts.[63]

Features of harness and saddlery suggest that the sculptors had a varying familiarity with horse-tack. Generally horses are provided on the head with headpiece, brow-band, throat-lash, cheek-strap, noseband and reins. A breast-strap and a breacher-strap cross chest and hindquarters. The saddle is usually covered by a fringed over-blanket but the shapes of covered saddle-horns are distinct.[64] Beyond this basic provision extra features sometimes occur, such as pendants, hanging pendant-straps, double shoulder-straps and narrow, 'triplet' straps (see Figure 6.6 above). Scenes XXXVII and LXXXIX are particularly well furnished, the former also having swords depicted in uniquely accurate fashion.

These are details that denote empirical observation of ridden horses, and they are paralleled on cavalry gravestones and in the artefactual record.[65] Nevertheless, there are some serious mistakes in practical detail. Breast- and breacher-straps should angle towards the saddle-pommels but sometimes they are rendered horizontally, revealing a clear ignorance of their function.[66] This is a common fault on gravestones.[67] Rather more strangely, frieze sculptors impossibly juxtaposed veristic details, such as the 'triplet' straps which hang down from the pommels of saddles when the latter are hidden by an over-blanket.[68] Dacian horses are much more schematically equipped, sometimes being shown barebacked, saddleless or with no saddle-horns, and with non-functional breacher-straps (XXVII, XXXI, CXLIV). The sculptors were intelligent observers, but not horsemen.

The depiction of saddle-pommels on Trajan's Column, other metropolitan monuments and provincial gravestones sits well with the growing body of artefactual evidence for Roman saddles in the form of copper-based alloy horn-plates and leather covers. Roman saddles were derived from Celtic models and it has been demonstrated through reconstruction and experimentation that the horns gave a very firm seat to the rider, making the lack of stirrups before the sixth century AD largely academic.[69]

The two vertically conjoining scenes XXXI and XXXVII depict scale-armoured barbarians on scale-armoured horses and these are unhesitatingly identified as Sarmatians by modern observers.[70] The armour hugs the horses' bodies closely, extending right down the legs to their hooves. While it is possible to protect human limbs with scale defences, there are problems with equine leg coverage. Indeed, hardly ever in military history have horses' legs been protected, and then only the shins, not the upper legs, were covered.[71] The horses should have been shown with blanket-like scale trappers such as those found at Dura-Europos in Syria, or depicted in various artworks.[72] Despite this misunderstanding, the column figures may be regarded as evidence for armoured Sarmatians on fully armoured horses, perhaps based on a literary description similar to that written by Tacitus, conceivably even one in Trajan's *commentarii*.[73]

There is one group of fourteen cavalry in the Roman forces which stands out from all others (LXIV). These men are universally identified as the Moors of Lusius Quietus.[74] They ride bareback with rope halters and no other harness or saddlery.[75] The horses themselves are indistinguishable in build or other features from the normal Roman mounts on the frieze (the comments on size above apply). This is a very successful depiction of Moorish light cavalry in ethnic dress, equipment and hairstyle, well corroborated by other iconographic sources.[76] The horses would have been hardy ponies, particularly suited to operation in difficult terrain.[77] Moorish cavalry were a valuable asset in the Danubian theatre, especially, perhaps, for scouting and foraging.[78]

Marching Roman infantry would have carried clothing and personal belongings, eating utensils, and perhaps some rations.[79] However, heavier items for communal, particularly contubernal use, such as milling-stones and tents, required pack-animals.[80] The sparse provision of carts only to selected march/camp scenes on the column is obviously artificial, although their loads are quite believable: wine, bulk items (tents?) and military equipment, including artillery. Clearly no real attempt was made by the sculptors to suggest the great numbers of pack-animals and vehicles which would have accompanied Trajan's armies on the march into Dacia. If just one animal served for every *contubernium* then a minimum of 640 mules formed the train of each Trajanic *legio*. To these must be added the spare animals necessary to replace losses and rotate the transport pool to reduce deaths through exhaustion. Then there was the wheeled train with a minimum of two beasts per vehicle. According to Vegetius, each *centuria* had an artillery piece of the type seen on the frieze (*ballista*) to his total of fifty-five.[81] Literally one per century would have involved 64 carts with 128 animals. Heavier artillery, if used in the Dacian wars, would have been disassembled and carried on wagons (10? = 40 animals?), but there would not have been the need for a great siege-train.[82] Other pack-animals, carts and wagons would have carried officers' tents and baggage, essential paperwork, religious ceremonial equipment, field-forges, tools, spare armour and weaponry. Thus a whole *legio* on campaign would probably have been accompanied by at least a thousand pack- and traction-animals, much more if beasts for senior officers, *centuriones*, *equites legionis* and cavalry remounts are factored in.[83] In addition there would have been the beasts accompanying the tail of slaves, servants, *lixae* and *calones* which followed the soldiers.[84]

In this context there is the problem of how much food for the Mediterranean dietary regime of the troops and fodder for the animals were carried with Trajan's armies as they advanced.[85] It might be assumed that grain from harvests the year before the campaign was strategically stockpiled at forward bases, notably legionary fortresses supplied by river (see below), ready to be transported with the expeditionary army. This would allow an early start to the war before the first grain-harvest, thus pre-empting barbarian concentrations of forces more dependent on that season's growth cycle for mass sustenance. This is how well-organised armies of the early modern period were given a boost forward to the discomfort of less prepared opponents: for example, the use of pre-stocked magazines by Ottoman forces (*menzil-hane*) and by the armies of Louis XIV.[86] A similar system of supply is attested for the Roman army.[87] Rather than attaching supply trains of carts and pack-animal strings to individual troop formations, they were pooled for the whole marching army, *legiones*, legionary *vexillationes*, *auxilia*, irregular units and allied forces.[88] Grain would have been issued down through the organisational hierarchy to *contubernium* level where it was milled into flour, which was baked into bread.

Wine, the main liquid staple, was also carried with the army, not in heavy and bulky *amphorae* but in barrels, as the column indicates.[89] In normal Roman practice the wine would probably have been mixed with locally available water, thus doubling the volume of liquid carried.[90] Meat on the hoof, gathered by tax in kind and requisition, would have transported itself until butchered.[91]

As the campaign unfolded and the season progressed, the army would shift increasingly to foraging as a source for grain.[92] The lands north of the Danube, particularly those within the Carpathian Ring, would have supported Roman forces, which in turn would have deprived the enemy of food. Scene CX on the column depicts citizen troops reaping at this stage of the season's advance. As Erdkamp observed, the men are cutting high up to take just the ears, not low down to take the stalks as well for fodder. In contrast, whole stalks are being cut on a relief from an unknown Trajanic(?) monument in Rome.[93] Sickles were normally part of the soldiers' equipment.[94] A number of literary accounts mention foraging on campaign, principally because troops were vulnerable to attack when disbursed in this activity.[95]

Fodder for animals was a different problem. Dry fodder was used to feed up transport animals and cavalry mounts after the lean winter period. However, the vast bulk and weight needed in the field, amounting to ten times the weight of human food, meant that in all periods it had to be gathered locally.[96] This more than the corn-growing season limited the start of campaigns to late April or early May when Ottoman armies campaigned during the sixteenth and seventeenth centuries in the same region as Trajan's forces.[97] Granted, Turkish troops were predominantly mounted, but Roman infantry still depended upon large numbers of pack- and traction-animals. Moreover, the area needed to support animals with grazing or reaped fodder was much larger than that which could supply troops with corn, thus the area of operations was much increased, and, again, dismounted cavalry in particular were vulnerable at this stage.[98]

It would be hopelessly speculative to play the numbers game further with pack- and traction-animals. The composition of Trajan's forces is not known with great precision, nor are the size and needs of the imperial entourage which included far more than just servants and secretaries.[99] Putative loads per head per animal species would probably have been kept below the theoretical maxima in an attempt to maintain workable condition.[100] Allowance would also have to be made by Roman planners for the losses inevitably sustained among beasts of burden on campaign through exhaustion, abuse and climate. These would not have been as horrendous as in the Mesopotamian theatre, but still posed problems of shortage.[101] The collection of many thousands of animals by provincial requisition would have formed part of the campaign preparations,[102] and it is quite likely that trade for horses in particular with the trans-Danubian barbarians preceded hostilities. It would have been impossible, and perhaps not even desirable, to keep news of planned mobilisation from Decebalus.[103]

From the foregoing examination several observations may be made about the frieze. There are too few wheeled vehicles and pack-animals for the realistic representation of armies on campaign. Servants, porters, *calones* and other non-combatants, particularly female followers, are almost entirely absent.[104] There are no camels. When it is depicted, baggage, with one exception (CVII), is only associated with citizen troops. Vehicles are light and the heavier, four-wheeled wagon is confined to barbarian use. With one questionable exception (LXVI), artillery is also the preserve of citizen troops and it is absent from sieges.

With regard to the routes taken into Dacia by Trajan's armies, scholars have advocated every possible line of advance that the unchanging geography of mountains and passes will allow.[105] This has proved to be an inconclusive exercise precisely because the scenery on Trajan's Column is almost always topographically unspecific. It is not evident how the sculptors could have been more 'helpful' using essentially two-dimensional conventions, and it is doubtful that this was ever their intention.

The itinerary in the surviving phrase of Trajan's *commentarii* provides a trajectory for the first war, taking the army via Berzobis and Aizi to the Iron Gates Pass and Tapae.[106] This is surely the route taken in both Trajanic wars and in the previous Domitianic advance under Tettius Iulianus,[107] leading as it did into the Dacian heartland. Viminacium would have been the main operational base on the Danube with magazines of *matériel* assembled by road and river before the wars started.[108] Berzobis would have been the next major base. Thereafter the site later occupied by Colonia Ulpia Traiana Sarmizegethusa may have formed the principal tactical base for operations against Decebalus' fortresses in the mountains.[109] These putative bases were spaced within the practical range of supply trains shuttling between them.[110]

Previous diplomatic missions and Domitianic campaigns in the region would have provided knowledge of the route. Itineraries of stopping-points and distances between them should have been available.[111] Pictorial 'maps' would have been neither practical nor necessary.[112]

TRAVEL AND TRANSPORT ON WATER

For water transport there are three main groups of scenes which figure boats and ships. All the vessels are drastically scaled down in size while humans and animals on board are at approximately the same scale (or slightly smaller) as in other scenes.

Right at the beginning of the helix the stage is set for the first great bridge-crossing and riverbank scenes (I–III). Of three ships with up-swept prows and sterns, steering-oars (in two cases) and cabins, two are loaded with barrels and one with trussed bundles (Figure 6.7). This type of vessel is also used to support the pontoon bridges figured in scenes IV and XLVIII. The scene is remarkably similar to one representing vessels moving upriver on the base of the Louvre Tiber God statue.[113]

In scenes XXXIII–XXXV Trajan appears to travel by ship along a river which is presumably the Danube (Figure 6.8). A flotilla of seven vessels is depicted. Two are of exactly the same type as in scenes I–III and are loaded with trussed bundles and shields as cargo. A third is similar, but has two oarsmen near the bow. A simplified hull with one oarsman unusually near the bow

Figure 6.7 Trajan's Column, scene II: river boat being loaded with barrels. Photograph J.C.N. Coulston

Figure 6.8 Trajan's Column, scenes XXXIII–XXXIV: biremes and horse-transport. Photograph J.C.N. Coulston

transports four horses as cargo.[114] Three very different ships are identified by their large crews, rams, steering-oars, sweeping stern-finials, outriggers and two banks of oars as bireme warships. Two have large, curving stern-cabins; one has a prow richly decorated with a triton and *putti*, and a swan figurehead (*cheniscus*).[115] Again Trajan travels in scenes XLVI–XLVII with a bireme and a cargo boat.

Between the two wars there is a similar, but much longer, imperial journey by water, interrupted by a series of landfalls. The presence of dolphins suggests a sea location (LXXIX).[116] In LXXIX three warships are rowed in line abreast. Their prows are decorated with sea-creatures and ships, and their cabins are all differently detailed (plain panels, textile, decorated panels), two have a *cheniscus*.[117] The nearest has a spritsail, the furthest a lowered mast(?). The outer two galleys are biremes but the middle one is a trireme with three oar-banks, the only example of its class on the column. Next there are two biremes (LXXXII) which appear, to judge by the activities of the crews, to be moored in harbour. The nearest has a spritsail and a ram-headed, above-water ram. Both ships are radically foreshortened in order to fit them into a very narrow space. In scene LXXXVI a bireme with lowered mast(?) is truncated to show only its forward half. Three warship sterns beyond a port wall serve as shorthand for a moored flotilla. On balance, the last ship in the sequence is probably also a bireme although there are no oars displayed (LXXXVII). It differs from all the others in having a standing mast with furled sail, mast-stays, hanging ropes, a raised and catted anchor, and an extraordinarily wide steering-oar blade. The stern is decorated with a very graceful *cheniscus*. All the warships on the column seem to ride too high in the water with their rams exposed.[118]

Towards the end of the second war, quite incongruously, two citizen soldiers work on a pair of diminutive, sweeping hulls located within some form of log stockade, while a column of Dacians moves past ignoring them completely (CXXXIII).

These vessels display in some cases almost random detail applied for visual variety. They are also subject to scaling down because they constituted some of the largest individual objects depicted on the frieze.[119] Thus, there are four types of vessel depicted: a barge(?) for horse-transport; mastless cargo-vessels for transporting military supplies along the river; bireme galleys used both at sea and on the river; and one trireme warship at sea (LXXIX). The scaling down and stylisation of vessels makes further elaboration problematic.[120]

Horse-transports are seldom depicted in ancient art but a mosaic from Henschir Medeina (Tunisia) shows a vessel with sweeps carrying three animals, named a '*hippago*'.[121] On the other hand, the oared warships on the column are very well represented iconographically. The trireme is identifiable from other sources and itself has contributed to research on the classical Athenian *trieres*. It was the largest class present in the Mediterranean fleets under the empire.[122] The upper ram in LXXXIII may also be seen on a relief from the Palatine in Rome.[123] With all the galleys the sculptors may have run together vertical and horizontal features near the bows so that there are no clearly defined outriggers for the upper oar-banks. The manner in which rowers hold their oars seems to be realistic.[124] In scene XLVI the sculptor confusedly rendered the furthest-forward oar as projecting from an incorrect, intermediate level.[125] The anchor in LXXXVII and the steering-oars are well paralleled in the archaeological record.[126]

The biremes are probably to be identified with the fast scouting *liburna* used by the Danube fleets as their principal oared warship. It was also in service with the Mediterranean fleets for communications and the movement of important personnel.[127] These ships would have been available to the sculptors for empirical observation at Rome and Portus. Access up to the city was possible for all classes of oared ships.[128]

The freight craft on the column lack clear indications of how they were propelled. They may have had sweeps (as in XXXIV) or they could have been provided with removable masts for towline attachment. In either case they are similar to freighters represented in numerous artworks on the Tiber and other rivers.[129] Their loads correspond with what is carried by pack-animals and carts on land: barrels (III), military equipment (XXXIII) and trussed bundles (II, XXXIII–XXXIV, XLVII). Barrels also appear on vessels in other pictorial sources.[130] The early river scenes show the build-up of supplies at Roman Danubian bases as part of the 'push' given to campaign commencement. Of course, full advantage was taken of the Danube and its navigable tributaries which formed lines of supply behind advancing armies. The advantages in economy of water transport were well recognised by Roman planners, as now,[131] but the Roman forces were not tied to or limited by river bases when campaigning north of the Danube, especially with the development of a good road network (see below). However, the vessels on Trajan's Column were most likely based on craft seen on the Tiber, not directly on Danubian traffic.

TRAVEL INSTALLATIONS

Elements of the Roman infrastructure are graphically represented on the column. As the armies advance, new roads are cleared through forests with much felling of trees and dumping of metalling.[132] Sometimes the exact functions of the soldiers' tasks are obscure, but industry is always evident, in contrast with barbarian inattention and indiscipline.[133] Many small streams and low rivers would have been forded on campaign but only once on the frieze do Roman troops resort to fording a watercourse and getting wet (XXVI). In this case citizen troops cross in orderly fashion and carefully preserve their military equipment from immersion. In contrast, the one time the barbarians are shown crossing a river (XXXI) chaos and panic ensues, horses are lost and some men even appear to be drowning. The orderly fording of rivers was a recognised Roman battle tactic, notably employing Batavian cavalry, which served to overawe and outmanoeuvre barbarian adversaries.[134]

Bridges are quite numerously depicted and most are wooden-framed constructions, supported on pilings, with open-work ballustrades.[135] Like the other buildings used as 'scenery' on the frieze, they are scaled down, insubstantial and rendered with a peculiar perspective.[136] There are much more substantial pontoon bridges across the Danube, built to withstand the heavy traffic of the full invasion armies at the commencement of campaigns (IV–V, XLVIII). It is thus very curious that the third major incursion into *barbaricum* (CI), which follows the pattern of the others in passing under an arch next to a major Roman base, crosses over an insubstantial pile bridge.

The great bridge in scene XCIX is distinguished from all others on the column by its stone piers and wooden superstructure (Figure 6.9). This certainly represents the great Danube bridge designed by Apollodorus of Damascus, corresponding as it does with representations on commemorative coins and surviving ashlar-faced concrete piers at Turnu-Severin in Romania (Figure 6.10).[137] However, on the frieze even this construction is subject to stylising truncation, five piers being shown rather than the twenty of the original. This bridge was one of the wonders of Roman engineering. Dismantled by Hadrian, and thus having only a short life, it was still worthy of praise in the sixth century.[138]

A great advantage that the Roman army had over all its adversaries was the construction of bridges over even the greatest waterways. Caesar's bridging of the Rhine on two occasions was an engineering triumph.[139] This ability

Figure 6.9 Trajan's Column, scene XCIX: bridge with stone piers, wooden superstructure and triumphal arches. Cichorius 1896–1900

Figure 6.10 Reassembled ashlar pier facing from Trajan's Danube bridge, Turnu Severin, Romania. Photograph J.C.N. Coulston

facilitated direct routes of march and supply, and perhaps lengthened the campaigning season. In the Danubian theatre this was particularly important, given the many rivers and wetland areas across Roman lines of advance.[140] Pontoon bridges were an efficient temporary means of crossing wide rivers employing available river craft.[141] Major Roman military pontoon constructions were thrown across the Rhine, Danube and Euphrates, and used for a majestic propaganda exhibition by Gaius across the Gulf of Pozzuoli.[142] This ability particularly impressed Dio and he carefully explained the method of construction in connection with Avidius Cassius' Euphrates bridge (*c*. AD 164).[143] A fine bridge of boats is depicted on a mid-second-century mosaic from Ostia (Nile or Rhône?).[144] If the Marcus Column is not simply emulating Trajan's Column, then there were pontoon bridges thrown across the Danube and other rivers in the Marcomannic War.[145] Such constructions continued to be used through the third and fourth centuries during British, Rhenish and Danubian campaigns.[146]

After the late Roman period it was not until the sixteenth century that pontoon bridges were again regularly constructed on this scale in the West. Most notable was the fortified example built by the Duke of Parma in 1584–5 on the Scheldt to blockade Antwerp successfully.[147] Not until the later seventeenth-century wars of Louis XIV were armies regularly accompanied by pontoon-bridging trains which allowed not only strategic campaign movement but also grand tactical advances to battle.[148] In this respect, as in many others, the Ottoman Turks were ahead of the West.[149]

On the other hand, permanent bridges posed a risk to the empire, hence the early dismantlement of the Drobeta superstructure despite protection provided by a bridgehead fort. The few Rhenish and Danubian bridges were fortified on the barbarian bank.[150] In any case the long Danubian river frontage was vulnerable to barbarian attacks, particularly by steppe cavalry in winter. When the Danube was low before major snow-melt in Central Europe, and especially when it was frozen, trans-Danubian peoples might cross and cause mayhem. Moreover, in wintertime the Roman army was withdrawn to winter quarters so scouting, spying, forward observation, diplomacy and other information-gathering activities would all have been at a low ebb.[151]

River ports appear along the Danube but without wharves being depicted.[152] Harbour moles and/or wharves are supported on arcades during the inter-war imperial journey (LXXX, LXXXI, LXXXVI). Parallels to these are provided by the great mole at Pozzuoli, by harbour installations shown in frescoes from Bay of Naples sites, and, close to home for the column sculptors, by the Portus north-east mole.[153]

The first installations on the frieze are clearly located along the River Danube, and most commentators agree that this is the location of the first two ship-borne journeys (XXXIII–XXXV, XLVI–XLVII). There is some agreement also that the seaport city depicted in scene LXXIX is Ancona, based on the hilltop temple, harbour arch and night departure (denoted by torches and the sole depiction of a ship's lantern). Thus the episodic journey that follows is presumably a maritime itinerary across to, and along, the Illyrian coast, perhaps via Salona in Croatia (LXXXVI). However, without Trajan's *commentarii*, the frieze cannot be pressed for any firm information between Ancona(?) and the Danube. This is clearly demonstrated by the fact that scholars have advocated every possible route, and some very unlikely ones, but without new evidence the exact itinerary will never be known.[154]

TRAVEL, TRANSPORT AND THE PRESENTATION OF ROMAN VICTORY

It could simply be asserted that Roman armies on campaign marched with tents, carts, pack-animals, equipment and provisions along roads and over bridges. Consequently all these activities, objects and installations appear on Trajan's Column. However, this would ignore the circumstance that they had never before been depicted in Roman propaganda art and they were never again represented with such care. Some further explanation of their role on the column is necessary beyond mere campaign narrative.

In many respects the reliefs of Trajan's Column are all about travel: travel of emperor and army to win victories, explore new regions, meet new peoples and to collect the royal gold which went to pay for the massive building complex within which the column stood. A 'meta-narrative' combined actions and events which did take place according to the *commentarii*, actions that would have happened in any Roman campaign, and information abundantly available for empirical observation in the capital. All was couched in the language of Roman 'historical' art for the urban audience.[155]

On the column the army is depicted as the emperor's partner and it is the army's qualities which gain prominence alongside the emperor's virtues.[156] There are many elements on the frieze which were not usual in Roman victorious iconography, and which could easily have been omitted, such as military architecture and construction scenes. Their inclusion is one of the many extraordinary features of the work, but they formed a central part of the propaganda message that is expressed through the roles and functions of the figure types.

The citizen troops seldom fight, but stand in reserve. It is principally they who deal with wheeled transport and artillery (see Figure 6.5 above). Only citizen troops engage in construction work (see Figure 6.2 above). Occasionally they do enter combat, but seldom in open battle rather than in technical siege contexts.[157] In contrast auxiliary troops guard, escort, scout, skirmish and do battle. In other words, all the technical activities, and, by extension, technical achievements, were attributed to the citizen troops who made up the politically significant element of the army. Thus the columns which march into *barbaricum* are depicted as well organised and equipped (see Figure 6.1 above). The very attention to military equipment detail creates a stark contrast with the unarmoured barbarians. Indeed, all the Roman troops in the field are armoured, except for some exotic irregulars who make up less than 4 per cent of the emperor's forces overall.[158]

River-freight, carts and pack-animals bespeak orderly, prior preparation. Artillery represents both the technical superiority of Roman citizens and their heritage of Hellenistic military technology. As they advance citizen troops construct surveyed camps, apocryphally based on a Greek model, to be filled with ranked tent-lines.[159] Military installations might symbolically represent the spread of Roman architecture and modes of living. Troops took their meals at set intervals on command and this was a recognised feature of army life and of campaign discipline.[160] At various points on the frieze reference is made to eating and drinking. Water is collected and drunk by soldiers and horses (XIV, XXI, LXXIV). Grain is carefully measured out using a *modius* (CXXIV).[161] Wine is carried forward with the armies and this could be seen as playing to a *topos* which assumed that northern barbarians drank no wine but subsisted on water.[162] There is some tangible support for this view from the paucity of *amphora* finds on pre-Roman sites in central Dacia, but this may be purely coincidental.[163]

Bridges and roads endowed the army with freedom of movement throughout the barbarian lands, areas where, in the Roman view, no facilities existed before. Rivers and bridges of course occupied an important place in traditional Roman perceptions of ritual space and liminality, and were seen to bound the empire.[164] River deities observed human affairs with interest and could play an active part in war. If not correctly propitiated they could be hostile; but bridging rivers was also a form of victory won over them.[165] This is clearly denoted by the triumphal arches on the bridges of Trajan's Column (see Figure 6.9 above), the Marcus Column, the Ostia mosaic and Domitian's Volturnus bridge.[166] The pontoon bridge on the Antonine Portonaccio Sarcophagus with its parade of bound barbarian prisoners is clearly a piece of triumphal imagery in a funerary context.[167]

On Trajan's Column the Drobeta bridge scene forms a vertical axis with the Victory, Decebalus' death and treasure scenes on the north-west face of the shaft (LXXVIII, XCIX, CXXXVIII, CXLV). Likewise, the vertical correspondence of bridges, emulated by the Marcus Column, does not just present all the major Roman invasions to the viewer, but also forms a victory axis up the south face.[168] Bridges also appear on a Trajanic relief at the Villa Medici in Rome and on Trajan's Arch of Benevento.[169] In his letter to Caninius Rufus that outlined worthy Dacian war subjects to write about, Pliny the Younger highlighted the discovery of new rivers and bridge-building.[170]

A clear contrast is also drawn in the contemporary presentations of Domitian and Trajan with regard to rivers. Pliny asserted that the Rhine and Euphrates admired Trajan, but that the Rhine and Danube rejoiced at Roman humiliations under Domitian.[171] Significantly, the Equus Domitianus in the Forum Romanum commemorated German victories by depicting Domitian's horse trampling the Rhine.[172] This very much took the victory-over-river line. The monument would have been known to all and, ironically, it was especially notorious after it had been torn down.[173] It cannot have been accidental that the Danube was figured on Trajan's Column backing the Roman side. At a very visible juncture (III) the deity benevolently waves on the advance of Trajan's army (see Figure 6.1 above).[174]

The movements of Trajan and his armies are almost always from left to right following the helix ever upwards. This is not solely dictated by the counter-clockwise helix but conforms with an iconographic tradition traceable in Greek art from at least the archaic period whereby the victor advances from the viewer's left.[175] Conversely, the barbarians are seldom able to stand up to the Roman advance but fall back in confusion and die in droves. They are rarely shown engaged in construction work, and they live in fortresses, not cities. There is contrasting care taken in the depiction of Roman public buildings, notably amphitheatres.[176] Dacians are at work in scene LXVII clearing trees, although it is not at all clear what they are trying to achieve by this. Ordered Roman marching columns contrast sharply with the movements of Dacian crowds.[177] Roman standards were primarily employed by the sculptors to locate the emperor on the frieze, but also to signal movement, whereas barbarian *dracones* and *vexilla* are more randomly applied.[178] Dacians do a lot of lurking in forests.

When the barbarians do advance it is without order and they cross watercourses in complete confusion, a theme picked up in the Villa Medici relief.[179]

In Roman literature, as in the case of Boudicca, and in Roman art, as on the Adamklissi metopes, when barbarians on the move had heavy wagons the latter impeded the owners, forming a killing ground and contributing to crushing defeat.[180] This is probably the implication in scene XXXVIII. When the Dacians assault Roman installations they are disorganised (XXXII, XCIV–XCV, CXXXIV) and, unlike similarly engaged Roman forces (CXIII), they have not prepared ladders. They do not have the capacity to equip themselves with armour, except, rather illogically, that shown on trophies (LXXVIII and column pedestal reliefs) and worn by the exotic Sarmatian cavalry (XXXI, XXXVII).

In each war the action progresses from Roman riverbank into deep *barbaricum*. In the first conflict this ends in Decebalus' submission at the main Dacian fortress. The second war artificially had to fill the upper half of the shaft, arbitrarily divided at midpoint by the Victory (LXXVIII). However, the sculptors evidently had much less 'narrative' material to play with. Thus the sequence commences with the long Trajanic sea-journey, continues with barbarian counter-attacks, and only really gets under way in scene CI. Once the main barbarian fortress has been taken (again?), the army advances even further from the Roman frontier. Decebalus' treasure is captured, he is tracked down and his head is taken for display. Thereafter the helix is filled with more messy skirmishes, which end at the top with hitherto unknown and differently dressed barbarians being 'contacted'.

This exploration of new peoples was very important. In specific scenes the sculptors took unusual care to distinguish the identities of barbarians through dress, hairstyle and equipment.[181] This is very similar to the range of peoples displayed on the Adamklissi *tropaeum* metopes and crenellations[182] and new barbarian *topoi* were being formed.[183] It reflects the diplomatic and military achievement of meeting barbarians who had not hitherto heard the Roman name, bringing them in to submit and to worship the emperor's *imagines* on military standards.[184] Rehearsal of ethnonyms, tribunyms and the enumeration of submissive barbarian kings on such monuments as the *Res Gestae Divi Augusti*, the La Turbie (France) *tropaeum*, the Arcus Claudii in Rome and the Mausoleum of the Plautii at Ponte Lucano (Italy) closely echo the visual iconographic references.[185] In a very real sense geographical and ethnographic exploration formed an integral part of, and greatly enhanced, the glory of any campaign.

The technical skills and achievements of the Roman army as depicted on Trajan's Column both made possible and advertised a victory of a singularly Roman kind. Aided occasionally by their own patron deities,[186] the emperor and his army are victorious not only over the barbarian enemy and his fortresses, but also over the natural forces of rivers, forests and mountains, all with their potentially hostile *genii*. Roman bridges chain rivers with their cooperation, roads gird the mountains and force through the Carpathian passes. Forests are cleared so that the army may move inexorably forwards (and upwards) in ordered columns. The conflict is not simply between Romans and Dacians, Trajan and

Decebalus. It is also a contest between Romans and barbarians, *disciplina* and *perturbatio*, *labor* and *inertia*,[187] imperial virtues and enemy perfidy,[188] civilisation and untamed nature, urbanisation and wilderness. The depictions of transport and travel on the column not only gave structure to the Dacian wars but also fully contributed to these actual and symbolic conflicts.

ACKNOWLEDGEMENTS

The writer is very grateful to Prof. A. La Regina and Arch. G. Martines for their extension of every facility for his work on the column scaffolding. Very many people have helped in discussing specific points, notably Amanda Claridge, Lynne Lancaster, Christine Morris and Roger Rees. Colin Adams, Hazel Dodge and Greg Woolf read the manuscript and made very many helpful corrections and suggestions. However, any opinions, mistakes and omissions herein are the writer's responsibility alone.

NOTES

1 Settis (1988) 549–89; *Colonna* (1988) 21–7; Stoiculescu (1985).
2 Vulpe (1969); Strobel (1984) 19–20; Lepper and Frere (1988) 226–9.
3 For example, Gauer (1977); Scheiper (1982); Settis (1988); Davies (1997).
4 Cf. Couissin (1926) Pl. IV–V and Robinson (1975) Pl. I. Coulston (1989); Bishop and Coulston (1993) 21–2.
5 Rockwell (1981–3); (1993) 233–42; Coulston (1989); (1990a); Wilson-Jones (1993); Claridge (1993); Lancaster (1999).
6 Cichorius (1896–1900).
7 Lehmann-Hartleben (1926) 7, 11–108, 116–17.
8 Ibid. 111–14, 145; Gauer (1977) 45–8, Fig.1; Farinella (1981); Scheiper (1982) 252–4; Brilliant (1984) 94–100, 103–8; Hannestad (1986) 167; Coulston (1990a) 298–300; (forthcoming a); Ampolo (1995).
9 Cuirassed officers (including Trajan), citizen troops (*legionarii* and *praetoriani* not distinguished), auxiliary infantry, auxiliary cavalry, standard-bearers, musicians, unarmoured soldiers, archers, slingers, bare-chested irregulars, Moorish cavalry, Dacians, Germans, Sarmatians, sacrificial attendants and *lictores*, civilians. See Coulston (1989); (1990a).
10 Lepper and Frere (1988) 30–1. Cf. Coulston (1990a) n.73.
11 Ibid., 303 *contra* Lepper and Frere (1988) 18.
12 Gauer (1977) 76–7.
13 Monte *et al*. (1998).
14 Dio 68.14.4–5; A. Gell., *N.A.* 13.25.1–2; Makkay (1995); Packer (1997) 220–1.
15 IV–VII, XXI–XXII, XXVI, XXXVI, XLVIII–L, LV, LXII–LXIII, CI, CVI–CVIII, CXXXI.
16 Connolly (1981) 242, 310; Junkelmann (1986) 196–205; Fuentes (1991).

17 IV, XLVIII–XLIX, LXXXVI–LXXXVII, XCVIII, CI–CII.
18 Robinson (1975) 48–75; Bishop and Coulston (1993) 93.
19 Coulston (1988b) 9; Keppie and Arnold (1984) No. 90; Frenz (1992) No. 5.
20 Cf. Glodariu (1976) 35–7.
21 Sanzi di Mino (1998) Figs 139, 144; Levi (1947) Pl. LXXX.d; Brett (1947) Pl. 30; Rostovtzeff (1957) Pl. XLVIII; Fradier (1989) 75; Junkelmann (1990) Fig. 72; Freshfield (1921–2) Pls XIX, XXII.
22 Toynbee (1973) 185–92; Goldsworthy (1996) 289–90, 293–4; Laurence (1999) 123–35.
23 Tac., *Ann.* 13.35; 14.38; SHA, *Pesc. Nig.* 11.1; *Sev. Alex.* 51.1; *Tyr. Trig.* 16.1.
24 II, XXXIII–XXXV, XLVII, LXXIV–LXXV, CVII.
25 VIII, XII–XIII, XVIII, XXI, XXVIII, XLIII, LIII, LVI, LXI–LXII, LXVI, XCVIII, CII–CIV, CVII, CIX–CX, CXIII, CXXV, CXXVIII, CXLI, CXLVII.
26 For example, Caes., *B.C.* 3.13; Livy 37.39.2; Tac., *Ann.* 13.35, 14.38.
27 Driel-Murray (1991a); (1991b); Pedley and Winterbottom (1991) 251–307, Fig. 223; Bishop and Coulston (1993) 102.
28 McIntyre and Richmond (1934) Fig. 1; Junkelmann (1986) Fig. 12.
29 Driel-Murray (1991b) 118.
30 Cf. SHA, *Claud.* 14.3. See also Kissel (1995) 221–5.
31 Dio 56.21.3; Coulston (1985) 270–1; Marsden (1969) 63, 131, 139, 168.
32 Polyb. 6.27–32; Jos., *B.J.* 3.77–84; Hyg., *mun. castr.*; Veg., *Mil.* 3.8; Connolly (1981) 136–7; Johnson (1983) 166–7; Baatz (1985); Davison (1989) 214–15; Pitts and St Joseph (1985) 162–3; Morel (1991); Gilliver (1999) 81–7. Cf. Parker (1988) 78–9.
33 Chandler (1976) 231–2. Cf. Geisberg (1974) 645–7, 1196–7; Strauss (1975) 345, 1284; Ward (1991) 72–3, 144–5, 231. The cautious use of parallels from other periods of military history is advantageous not so much in order to demonstrate what was Roman practice as to place the Roman capabilities and achievements within a broader practical context. This is particularly true of geographical, transport and logistics discussions from which comparisons may be drawn without each period's scholars having to reinvent the wheel (Cf. Crefeld (1977); Bachrach (1985); Lynn (1993); Southern (1996)). Sixteenth–seventeenth-century Ottoman Turkish warfare in the Danube lands provides attractive comparanda for Roman activities in the same theatre. The Turks developed well-organised supply systems ahead of their Western enemies, using Roman roads, pack-camels and sophisticated bridging methods; they transferred military formations back and forth between European and Mesopotamian fronts. In the *kapu kulu* formations the Porte had a permanent, standing army to supply, deploy and plan for, the first such operating in Europe since the middle Byzantine period (Perjés (1970); Finkel (1988); Kortepeter (1973); Murphey (1999); Haldon (1999)). Against this the Ottoman forces were predominantly mounted, had to deal with a cumbersome gunpowder artillery siege-train and had a different dietary regime, notably lacking wine. They operated over an immense range with a journey of 1912 km from Istanbul to Vienna and back (Murphey (1999) Map 4). For Roman campaigning and supply on the Danube see Kissel (1995) 268–71.
34 Plut., *Luc.* 11.4, 34.3; Tac., *Ann.* 15.12; Jos., *B.J.* 4.436; SHA, *Sev. Alex.* 47; *Claudius* 14.3; Veg., *Mil.* 3.23; Amm. 28.6.5; *Not. Dig.*, *Or.* XXXI. 48, 54, 57;

XXXIV.33; *CIL* III, 93, 123; XVI, 106; Gilliam (1965) 76; Davies (1967) 116–17; (1969a) 227–8; (1969b) 432–5; Toynbee (1973) 138–9; Dabrowa (1991); Roth (1999) 207–8.

35 Bulliet (1975) 20–5, 87–110; Engels (1978) 14–15; Kuhnen (1991); Roth (1999) 207; Murphey (1999) 76.

36 Geisberg (1974) 261–3, 266, 1188, 1196–7, 1199, 1203, 1220, 1314; Strauss (1975) 1386; Finkel (1988) 91, 142; Murphey (1999) 70–7, 80–1, 101.

37 Strauss (1975) 33. Cf. Laubscher (1975) Pl. 28.3.

38 *Contra* Goldsworthy (1996) 293.

39 Wilkes (1983) n. 40; Coulston (1985) 297.

40 Strobel (1984) 65, 102–3, 118, 126, 136, 148–50.

41 Hyg., *mun. castr.* 29, Luff (1982) 265, Bökönyi (1988) Table 1.

42 Toynbee (1973) 137–40; Bulliet (1975) 167, Fig. 79; Knauer (1998) 24–6, Fig. 4.

43 XL 8 spokes, LXI 12, LXVI 8, CVI 6, CVII 12(?), CXXIX 8.

44 Burford (1960) 9–10; Perjés (1970) 10–11; Mitchell (1976) n.100; Engels (1978) 14–16; Buck (1983) 48–9; White (1984) 129–31; Goldsworthy (1996) 293–4; Junkelmann (1997) 60–2; Erdkamp (1998) 70–3; Roth (1999) 204–7; Murphey (1999) 77–8.

45 Hassel (1966) Pl. 20.1; Andreae (1979) Pl. 80.2; Brilliant (1967) Pls 44–6, 48); Petersen *et al.* (1896) scenes XVI XXV, XXVIII, XXXIX, LXIV, LXXXV, XCIII, CXI.

46 Ibid., scene XXV; Espérandieu (1907–81) No. 3232; Tchernia (1986) Fig. 5.1; Molin (1990) Figs 9–12, 14. Cf. Bulliet (1975) Fig. 94.

47 For example, Espérandieu (1907–81) Nos. 3245, 5499; 1931, No. 455; Krüger (1970) No. 330; Casson (1974) Pl. 13; Bender (1978) Figs 2, 11, 13.

48 Vulpe (1969) 218–22; Florescu (1965) 638, Figs 218–21.

49 Richmond (1935) 14; Marsden (1969) 189–90, Pls 6–8; 1971, 215, Figs 8–9; Baatz (1978) 9–16, Figs 9–11; Bishop and Coulston (1993) 166–7, Fig. 120.3.

50 Blyth (1992) 152.

51 Baatz (1966) 122–8; Marsden (1969) 123, 188–90, Pls 1, 3, 6; 1971, 208–33; Baatz (1978) 11–16; Schramm (1980) Fig. D, Pls 5, 9, 11; Baatz and Feugère (1981); James (1983) Fig. 4; Bishop and Coulston (1993) 81, 139, Figs 44.1, 99.1; Chevedden (1995) 135–6.

52 Couissin (1926) 452; Alfs (1941) 112; Robinson (1975) 180; Bishop and Coulston (1993) 85; Schlüter (1993) 119–20.

53 Baatz (1966) 195; (1994) 128–33; Marsden (1969) 180–4; Campbell (1986).

54 Caes., *B.G.* 7.25; Jos., *B.J.* 5.296; Zos. 1.70; Amm. 19.1.7; Procop., *Wars* 5.23.9–12. Cf. Richmond (1968) 32–3, Figs 14, 18.

55 James (1983); Bishop and Coulston (1993) 139, Fig. 99.8.

56 Veg., *Mil.* 2.25, 3.14, 3.24; Maur., *Strat.*12.B6; Chevedden (1995) 138–42, 152.

57 Marsden (1969) 181, 184–5, 193–5.

58 See Dio 68.9.3.

59 XII, XXXVIII, L, LXXII, LXXVII, CXXIX.

60 A Burus? Dio 68.8.1. Cf. Vulpe (1963) 224–6; Ampolo (1995).

61 Led: V, VII, XXI, XXXVI, XLII, XLIX, LXXIV, CIV. Ridden: XXI, XXIV, XXXVII–XXXVIII, XL, LVII–LVIII, LXXXIX, XCVII, CII, CXLII–CXVI.

62 For example, Schleiermacher (1984) Nos 4, 15, 20, 35, 42.
63 Luff (1982) 252–8; Davison (1989) 143–8; Junkelmann (1990) 32–44; Hyland (1990) 11–29.
64 Cf. Leander Touati (1987) Nos 20, 30, 40, 80.
65 Bishop (1988); Bishop and Coulston (1993) 105–7.
66 XXI, XXIV, LVII–LVIII, XCVII, CII, CIV, CXLV.
67 For example, Schleiermacher (1984) Nos 4, 35, 56–7, 67, 73–4, 77, 122.
68 V, VII, XXI, XXXVI–XXXVIII, XLIX, CIV.
69 Coulston (1986) 61–2; Connolly (1987); Hyland (1990) 131–6; Junkelmann (1992) 34–74; Bishop and Coulston (1993) 105–7.
70 For instance, by Cichorius (1896–1900) II, 150, 179–84; Richmond (1935) 17; Patsch (1937) 22–3; Gauer (1977) 26–8; Strobel (1984) 31, 178, 181; Lepper and Frere (1988) 80, 85; Wilkes (1983) 271–2; Mielczarek (1993) 34–5.
71 Hel., *Aeth.* 9.15, perhaps fanciful, but see modern police horses on riot control.
72 Robinson (1975) Figs 8, 65, 81–2, Pl. XXVI; Gall (1990) Pls 3–4, 9–11; Mielczarek (1993) Figs 6, 9, 11; Bishop and Coulston (1993) Figs 113–14.
73 Tac., *Hist.* 1.79; Coulston (forthcoming b).
74 Dio 68.8.3; Speidel (1975) 212–14; Strobel (1984) 68–71, 151–2, 195–6; Lepper and Frere (1988) 91, 94, 105, 276.
75 Cf. Strabo 17.3.7.
76 Rostovtzeff (1946) Pls XI–XII; Horn and Rüger (1979) 580, 640–1; Connolly (1981) 148; Franken (1999).
77 Hyland (1990) 11–14.
78 Glodariu (1974). Cf. Murphey (1999) 67–8 for early modern Tatar cavalry on steppe horses operating in the same region.
79 Kromayer and Veith (1923) 423–6; Watson (1969) 62–6; Junkelmann (1986) 196–207; (1997) 86–93; Erdkamp (1998) 76–80; Roth (1999) 68–77.
80 Junkelmann (1986) 207–11; (1997) 110–27; Pitts and St Joseph (1985) 181–2; Erdkamp (1998) 35–6; Roth (1999) 77–80.
81 Veg., *Mil.* 2.25. Cf. Jos, *B.J.* 3.166.
82 Goldsworthy (1996) 289. Cf. Veg., *Mil.* 2.25.
83 SHA, *Claud.* 14.3; Kromayer and Veith (1923) 394–5, 500, 528–9; Goldsworthy (1996) 289–90; Roth (1999) 82–3; Laurence (1999) 127–9.
84 Speidel (1989); Goldsworthy (1996) 72–3, 295–6; Erdkamp (1998) 41–2; Roth (1999) 91–115; Gilliver (1999) 29–31. Cf. Geisberg (1974) 250, 1189.
85 Watson (1969) 63–4; Davies (1971); Junkelmann (1986) 123–7; (1997) 103–9, 150–1, 176–81; Goldsworthy (1996) 291–2; Erdkamp (1998) 27–34; Roth (1999) 15–44; King (1999).
86 Perjés (1970) 19–25; Crefeld (1977) 17; Finkel (1988) 151–6; Lynn (1999) 55, 112–13, 137, 145; Murphey (1999) 25, 70–1, 97–9.
87 Anderson (1992) 35; Erdkamp (1998) 46–52; Roth (1999) 169–77; Gilliver (1999) 58–60.
88 Polyb. 6.40; Caes., *B.G.* 2.19; Jos., *B.J.* 3.124, 5.47–9, 132; Tac., *Ann.* 1.51, 13.40; Arr., *Alan.* 8; Veg., *Mil.* 3.6; Kromayer and Veith (1923) 546–8; Peddie (1994) 42–58; Goldsworthy (1996) 287–8; Erdkamp (1998) 73–4; Roth (1999) 79–91; Gilliver (1999) 39–45.
89 Bender (1978) Fig.13.

90 Tchernia (1986) 13–19; Roth (1999) 119–23.
91 Erdkamp (1998) 31–3; Roth (1999) 28–31. Cf. Kirk (1979) 44; Finkel (1988) 173–89; Murphey (1999) 88–90.
92 Cf. Crefeld (1977) 7–39; Goldsworthy (1996) 290–1; Erdkamp (1998) 122–40; Roth (1999) 130–4; Gilliver (1999) 60–1.
93 Erdkamp (1998) 124; Giuliano (1981) No. III.9; Koeppel (1985) No. 1.
94 Jos., *B.J.* 3.95; SHA, *Claud.* 14.6.
95 For example, Caes., *B.G.* 4.32.
96 Perjés (1970) 15; Crefeld (1977) 23–6, 29, 34–5, 38; Finkel (1988) 193–7; Murphey (1999) 22–5; Lynn (1999) 54, 78–9.
97 Perjés (1970) 14–7; Finkel (1988) 193–7; Murphey (1999) 20–1.
98 For example, Jos., *B.J.* 6.153; Cf. Crefeld (1977) 35; Erdkamp (1998) 125–30; Roth (1999) 286–92; Gilliver (1999) 60.
99 Scarborough (1985); Lepper and Frere (1988) 89; Austin and Rankov (1995) 114.
100 Murphey (1999) 76–7.
101 Cf. Crefeld (1977) 12; Kirk (1979) 45–6; Erdkamp (1998) 80; Roth (1999) 65, 203; Murphey (1999) 69.
102 Tac., *Ann.* 2.5. See Kissel (1995) 234–7.
103 Hyland (1990) 22–3; Aymard (1951) 177–9, 524; Anderson (1985) 105–7; Bökönyi (1988); Austin and Rankov (1995) 26–7.
104 Boymel Kampen (1995) 53.
105 For example, Davies (1917); (1920); Patsch (1937) 62–4, 70–84, 106–7; Gauer (1977) 13–25, 28–31, 36–7, Fig. 2; Cisek (1983) 296–300, 323–5; Lepper and Frere (1988) 39–41.
106 Ibid. 227–8.
107 Cf. Dio 67.10.2; 68.8.1.
108 Roth (1999) 165–89.
109 *Atlas* (1974–9) Map VII.1.1, 3; Lepper and Frere (1988) Map III.
110 Cf. Erdkamp (1998) 73; Roth (1999) 200.
111 Patsch (1937) 3–27; Southern (1997) 92–100; Sherk (1974) 558–60; Millar (1982) 15–18; Dilke (1985) 112–29; (1987) 236–8; Syme (1988) 234–5; Nicolet (1991) 85–94; Austin and Rankov (1995) 37–8, 112–18; Mattern (1999) 27–9. Cf. Parker (1972) 83–6, 102–5.
112 *Contra* Koeppel (1980).
113 Lembke (1994) Pls 23–5. Cf. Tchernia (1986) Fig. 2.
114 Cf. Casson (1994) Fig. 98.
115 App., *Met.* 11.
116 Cf. Becatti (1961) No.106.
117 Basch (1987) Figs 989–95, 998–1001.
118 A common artistic practice (Reddé (1986) Figs 31, 34–6, 41, 44, 54; Casson (1994) Figs 54, 71, 74).
119 Basch (1987) 445–6.
120 Zahariade and Bounegru (1994).
121 Reddé (1986) 397; Casey (1994) 154, Fig.16; Viereck (1996) 85, Fig. 76.
122 Starr (1941) 53–5, 108; Morrison and Coates (1986); Reddé (1986) 110–12; Casson (1994) 60–77, 94; Gardiner (1995) 64–6, 81, 137–40; Viereck (1996) 37–44.

123 Reddé (1986) Figs 54–7; Tomei (1997) No. 39. Cf. Reddé (1986) Figs 36, 44; Gardiner (1995) 68; Viereck (1996) Pls 15, 31 (as distinct from a simple figurehead, Reddé (1986) Figs 49–53).

124 Morrison and Coates (1986) xvi.

125 Basch (1987) Fig. 986.6.

126 White (1984) 151–2; Casson (1994) 119–20, Figs 91–2; Viereck (1996) 132–5, Fig. 62, 124–9; Schalles (1988).

127 Starr (1941) 19, 34, 54–5; 108; Sarnowski and Trynkowski (1983); Morrison and Coates (1986) 8–9; Reddé (1986) 104–10; Zahariade and Bounegru (1994) 34–7; Casson (1994) 94–5; Gardiner (1995) 72–3, 81–2, 88–90, 140–1; Höckmann (1995) 83; Austin and Rankov (1995) 177–8; Viereck (1996) 35–7; Baatz and Bockius (1997) 10–15, 32–5, 41–2, 51–3.

128 Dion. Hal. 3.44.3; Virg., *Aen.* 8.81–93; Suet., *Gaius* 15.1; Casson (1965) 32; Meiggs (1973) 291.

129 Casson (1965) 1994, 127–34; Meiggs (1973) 293–8; Zahariade and Bounegru (1994) 37–9; Viereck (1996) 84, Pls 51–2.

130 For example, Espérandieu (1907–81) Nos 4072, 5184, 5193, 6699.

131 White (1984) 131–2; Greene (1986) 17–18, 34, 39–40; Anderson (1992) 58–70. See Starr (1941) 124–66; Reddé (1986) 356–62; Kissel (1995) 268–71; Austin and Rankov (1995) 177–9, 203.

132 XXIII, LVI–LVII, LXIX, XCII, XCVII.

133 Coulston (1990b) 42–6; Mattern (1999) 203–5.

134 Dio 60.20.2, 69.9.6; *CIL* III.3672 = *I.L.S.* 2558; Veg., *Mil.* 3.7; Speidel (1994) 45–7; Austin and Rankov (1995) 174.

135 XII, XIV, XV, XVII, XIX(?), XXI, L, LVI(?), LVIII, XCII, CI, CVII(?), CXXXI.

136 Saatmann *et al.* (1939) 189–99; O'Connor (1993) 133–9.

137 Barcacila (1966); Belloni (1973) Nos 355–9, Pl. 17; Coulston (1990b) 49; O'Connor (1993) 142–5.

138 Dio 68.13.1–6; Procop., *Aed.* 4.6.11–16.

139 Caes., *B.G.* 4.17–18, 6.29; Flor. 1.45.15 Saatmann *et al.* (1939); Connolly (1981) 240–2.

140 *Atlas* (1974–9) Map V.1; Murphey (1999) 68–9, Map 5.

141 Flor. 1.45.14. Cf. Tac., *Hist.* 5.19.

142 Tac., *Ann.* 15.9; Suet., *Gaius* 19.

143 Dio 71.3.

144 Becatti (1961) No.108; Dilke (1987) 246–7, Fig. 14.7; Saatmann *et al.* (1939) 199–202.

145 Petersen *et al.* (1896) scenes III, XXIX–XXX, LXXVIII, LXXXIV, CVIII.

146 Veg., *Mil.* 3.7; *B.M.C.* V, 353; VI, 209, No. 967; Austin and Rankov (1995) 175.

147 Strauss (1975) 575, 669; Parker (1977) 214–15. The Spanish Army of Flanders was the first real professional army in the West after the late Roman period. Long individual service, (ir)regular pay scales, standing *tercio* formations, long-distance supply and itineraries along the Spanish Road distinguish it from earlier armies (Parker (1972)).

148 Propaganda surrounding Louis XIV's pontoon bridge over the Rhine constructed in 1672 during the Dutch War (Lynn (1999) 114–15) very

consciously echoed Caesar's achievements on the same river. Louis's direct interest in Trajan's Column led to his commissioning both the first complete set of casts and the first overall drawing of the frieze (Settis (1988) 591–3; *Colonna* (1988) 27–30). The latter (Bartoli (1672)) was published in the same year as the Rhine bridge was built, and a link between contemporary bridge and column bridges would surely have been made in the court of '*Il Traiano di Francia*'.

149 Murphey (1999) 24, 140. See also Marshall (1993) 179. For the west see Parker (1972) 82; Strauss (1975) 1284; Lynn (1999) 114–15, 140–1, 155, 207, 218–19, 288, 320, 353; Chandler (1976) 228–30; Ward (1991) 286–7, 304.

150 Austin and Rankov (1995) 174–7.

151 Ibid. 173–4; Coulston (forthcoming b).

152 II–III, XXXIII, XXXV, XLVII.

153 Meiggs (1973) 158, Pl. XVIIIa; Clairmont (1975) 32; Blackman (1982a) Fig.1B, 5; (1982b) Fig.7.

154 Cichorius (1896–1900) III, Fig. 6; Stuart Jones (1910) 444–6; Patsch (1937) 95–6; Stucchi (1960) Fig.1; Clairmont (1975) 26–35; Gauer (1977) 32; Lepper and Frere (1988) 129–39.

155 Hannestad (1979); Koeppel (1982); Hölscher (1980) 290–7; Baumer *et al.* (1991); D'Ambra (1998) 80–91.

156 Richmond (1935) 2; Coulston (forthcoming a).

157 Sculptors mistakenly show auxiliaries building in scenes XII and CXXIX. The first instance is early in the project when some carvers were still unsure of figure-type parameters and the second is in a scene with other clear confusions and mistakes. Altogether only three auxiliaries build, compared with 179 citizen troops.

Citizen troops in battle: XL, LXXII. The first is a hard-fought contest, the only one in which Romans appear wounded. The second is a confused scene where figure-type features are applied in mistaken combinations. Citizen troops in sieges: LXXI (*testudo*), XCVI, CXIII, CXV–CXVI.

158 Ten slingers (LXVI, LXX, LXXII, CVIII, CXIII), 25 bare-chested irregulars (XXIV, XXXVI, XXXVIII, XL, XLII, LXVI, LXX, LXXII, CVIII, CXV), 14 Moors (LXIV).

159 Gilliver (1999) 66–9.

160 Jos., *B.J.* 3.86. Cf. Tac., *Hist.* 4.39.

161 Cf. Becatti (1961) Nos 58, 87, 89, 98, 115, 133, 135; *R.I.B.* II, 2415.56.

162 Pliny, *N.H.* 16.4; Tac., *Ger.* 23.1; Dio 49.36.2; Mattern (1999) 72.

163 Glodariu (1976) 10–17. Cf. Villard (1990); Woolf (1998) 174–80.

164 Alföldi (1952); Millar (1982) 19–20; Whittaker (1994) 75–8, 170–1; Braund (1996).

165 Plut., *Luc.* 24.7; Dio 40.18.5; Arr., *Perip.* 10; Tac., *Hist.* 5.17; Flor. 1.45.15; 2.13.88; Stat., *Silv.* 4.3.81–4; Kleiner (1991).

166 Petersen *et al.* (1896) scenes III, LXXVIII; Stat., *Silv.* 4.3.95–100.

167 Bertinetti *et al.* (1985) 179.

168 IV, XLVIII, CI, CXXXI. Gauer (1977) Fig. 1.

169 Koeppel (1985) No. 13; Hassel (1966) Pl. 13.

170 Pliny, *Ep.* 8.4.2.

171 Pliny, *Pan.* 14.1; 82.4.

172 Stat., *Silv.* 1.1.50–1; Steinby (1995) 228–9, Figs 77–80; Darwell-Smith (1996) 227–33.
173 Suet., *Dom.* 23; Dio 68.1.1.
174 Cf. Virg., *Aen.* 8.57–9.
175 For example, Boardman (1978) Figs 212–13; (1985) Fig. 127; Smith (1991) Figs 195, 209.
176 II–III, XXXIII, LXXIX–LXXXI, LXXXVI. See Coulston (1990b) 48–9; Futrell (1997) 89–91.
177 XXIV, XL–XLII, LXIV, LXVI, CXXII, CXXXII–CXXXIII, CXXXVI, CXL.
178 Roman standards: IV, VII, XXII, XXVI, XLVIII, CVI, CVIII. Barbarian standards: XXIV, XXXI, LIX, LXIV, LXVI, CXXII.
179 Koeppel (1985) No. 13.
180 Tac., *Ann.* 14.37: Dio 62.12.5; Caes., *B.G.* 1.26, 4.14, 2.51; Florescu (1965) Figs 218–19, 221.
181 Germans, Sarmatians, northern Dacians: XXVII, XXXI, XXXVII–XXXVIII, CLI.
182 Florescu (1965) Figs 184–5, 195–206, 213–21, 228–31, 233–58.
183 Bülow (1980); Demougeot (1984); Mattern (1999) 70–80.
184 Suet., *Gaius* 14.3; Tac., *Ann.* 15.29; *I.L.S.* 986. Cf. Pliny, *Pan.* 82.4.
185 Conole and Milns (1983); Nicolet (1991) 20–4, 32–4; Mattern (1999) 30–1, 40–1, 162–9.
186 III (Danubius), XXIV (Jupiter Tonans), XXXVIII, CL (Nymphs?).
187 Tac., *Ger.* 15.1; Mattern (1999) 204–5.
188 XXXI–II, LXXV–LXXVI and XCIII–XCV, CXXXIV.

7

'THERE AND BACK AGAIN'

Getting around in Roman Egypt

Colin Adams

The purpose of this chapter is to introduce the evidence for travel provided by the papyri of Roman Egypt.[1] It is not intended to be deeply analytical, rather it describes the type of evidence contained in these documents. The chapter is concerned with two main sources of evidence on papyri: first, private letters, which contain valuable information on how and why people travelled in Egypt, and what dangers or problems they might have encountered; and second, itineraries of specific journeys, which may be interesting if compared to larger and more important itineraries discussed elsewhere in this volume.

Before considering our evidence, however, it is necessary to outline the geography and transport infrastructure of Egypt, for any study of travel, transport or travel in a particular region must take account of geographical and topographical factors. The River Nile is the desert's only relief, and thus it was, as today, that habitation was only really feasible in the narrow river valley: at its widest point perhaps 10 km. This valley runs from Syene (modern Aswan) to Babylon (just north of modern Cairo), where it fans out into the Nile delta. On each side of this river valley lay desert: to the east, a mountainous desert reaching to the shores of the Red Sea; and to the west, the sandy expanse of the Sahara. This latter desert was interrupted only by a small number of oases – the Fayum depression, which lay close to the Nile valley, the Small and Great oases, and finally the famous oasis at Siwa, the site of the Oracle of Ammon, visited by Alexander the Great.

The River Nile provided the most natural of transport highways, and since the very beginning of Egyptian history had provided the main artery of communication in the country. Although transport by land was no doubt part of everyday life, most long journeys would have been made by river. Canals linked the Nile with other parts of the valley, and this was especially important in the Delta, where many canals formed navigable channels. Strabo notes the canals on the first stages of the journey south from Alexandria, leading first to Canopus and Canals on the Schedia, and then south towards Memphis.[2]

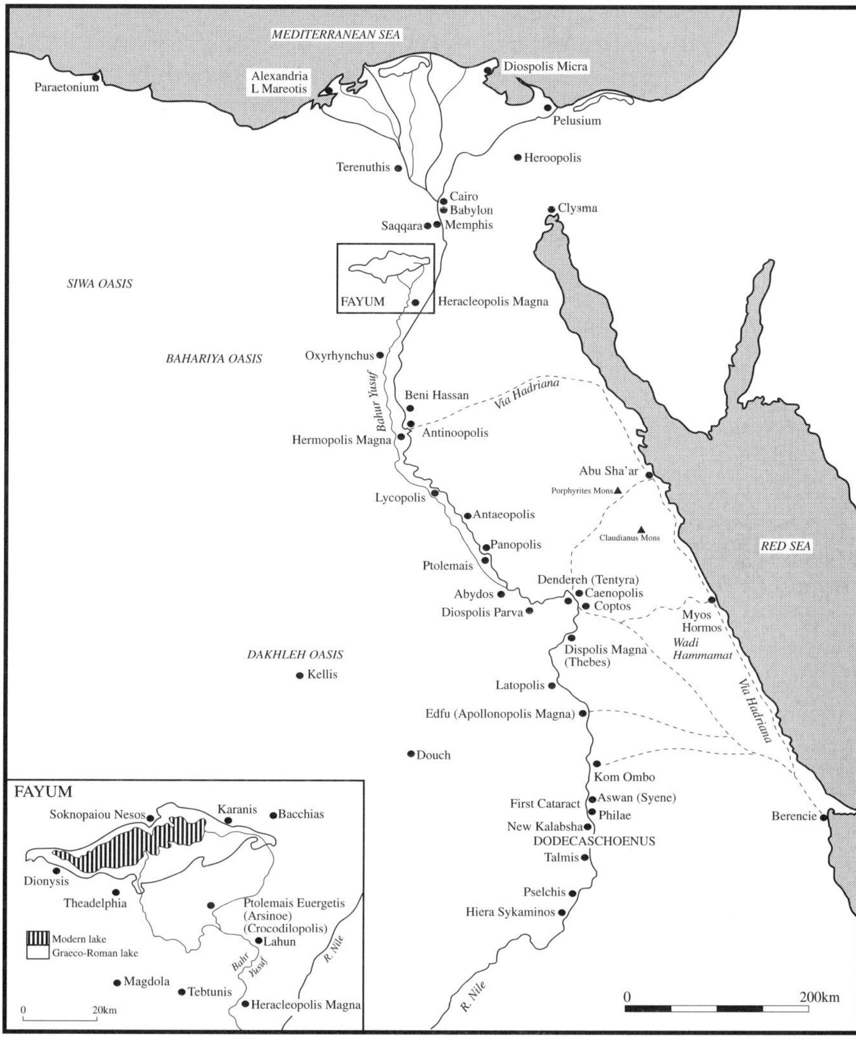

Figure 7.1 Egypt: the Nile and major routes to the Red Sea

By land, most people would have travelled by foot, as most journeys would have been short. Villages lay close together, and even the capitals of the regional districts of Egypt, known as *nomes*, would have been a day's walk away at most. This is well illustrated by a letter from the Roman period, exchanged between friends living in the villages of Karanis and Bacchias in the Fayum, where we read: 'Theon my brother salutes you and urges you to come to us from Bacchias. For observe that those who come from there arrive within the second hour'.[3] So short journeys by foot could be easily made; perhaps the only significant obstacle would have been the topography of the surrounding land, which was dominated by many irrigation channels and dykes.[4] However, roads and tracks often followed these irrigation channels, or lay along the tops of dykes, and thus provided well-trodden straight routes, probably well supplied with basic bridges – just as in rural Egypt today. These routes, however, would only have suited travel by foot or on pack-animals; they would certainly not have been suited to wheeled vehicles, apart from small carts. One private letter from Oxyrhynchus illustrates nicely the fact that short journeys could be made either by canal or by donkey:

> Greetings, my lady Serenia, from Petosiris. Make every effort, lady, to come out on the 20th for the birthday of the god, and let me know whether you are coming out by boat or by donkey, so that it can be sent for you. Take care not to forget, lady. I pray you are well and long remain so.[5]

Larger roads and thoroughfares certainly existed, but unfortunately, at least in the Nile valley, there is no archaeological record – agricultural activity and changes in course of the river have ensured this. It is certain, however, that major roads leading north and south stretched along the Nile valley roughly parallel to the course of the river. So much is clear from the Antonine Itineraries and the Peutinger Table. The roads connected the main *metropoleis* of the *nomes* from Syene (modern Aswan) to Babylon (near modern Cairo), where they split into two, one towards Alexandria, the other Pelusium. These routes existed for both state and civilian use. They were part of the system of empire-wide communication, the imperial post, which, according to Suetonius, was instituted by Augustus, but which was probably based on earlier state communication systems such as that mentioned by Herodotus for Persia, and that which the Ptolemies created in Egypt.[6] There is no reason to suspect, however, that the routes were not used by civilians, but it is reasonable to assume that charges were made.

This was certainly the case in the Eastern Desert, and we are lucky in that we have a fair amount of literary, papyrological and epigraphic evidence, coupled with an ever-increasing body of archaeological detail.[7] Important routes traversed the Eastern Desert from Koptos, the main Nile emporium, and Qena in the Nile valley, to Berenike, Myos Hormos, Mons Claudianus and Mons Porphyrites (running south to north).[8] The emperor Hadrian ordered the building of a new route, the via Hadriana, which led from Antinoopolis to Berenike. An inscription relating to its construction by teams of soldiers describes the nature of the route, which must be representative generally of the routes traversing the desert:

> Imperator Caesar Traianus Hadrianus Augustus, son of the divine Traianus Parthicus and grandson of the divine Nerva, pontifex

> maximus, holding the tribunician power for the twenty-first time, the title of imperator for the second time, the consulship for the third time, father of his country, built from Berenike to Antinoopolis through secure and level country to the Red Sea the new Hadrian way spaced with abundant cisterns, resting stations, and garrisons. The 21st Year. Phamenoth 1. [Names of soldiers follow].[9]

Cisterns and resting stations (*hydreumata*) seem to have been an innovation of the Roman period, as Strabo records that before the time of Augustus

> the camel merchants travelled by night using the stars for guidance, like the sailors. They carried water with them when they travelled. However, now they have built watering stations by digging down to a great depth and they have constructed cisterns for rain water although it is scarce. The journey takes six or seven days.[10]

The reason behind this development is implied by Strabo, who notes the great increase in the amount of trade between the Roman empire of the east, through the Red Sea ports – good road facilities were therefore important.[11] The provision of these facilities has important implications for our understanding of the Roman economy, given the government's interest in trade and commerce in this region.

Few milestones exist from Egypt, although some were reported by Petrie in the Western Desert north of the Fayum, and may have marked a route from there north to Saqqara.[12] He found milestones marking the route, a tablet at each *schoenus* (four miles) and a smaller upright block every third of a mile (1,000 cubits). In the Eastern Desert, where routes are easier to detect and are more frequently attested in our sources, milestones have not been found. Marker cairns, however, are common, but seem to mark the edges of roads rather than denoting distance.[13] Strabo records an interesting detail when describing his journey in a carriage from Syene to Philae:

> Along the whole road on either side one could see in many places a stone like our *Hermae*; it was huge, round, quite smooth, nearly sphere-shaped, and consisted of the black, hard stone from which mortars are made – a smaller stone lying on a larger, and on that stone again another. Sometimes, however, it was only a single stone; and the largest was in diameter no less than twelve feet, though one and all were larger than half this measure.[14]

These stones must date to at least the Ptolemaic period, and possibly much earlier – a nineteenth-century traveller described these stones, noting that hieroglyphics had been carved into some of them.[15] It is difficult to believe that Strabo would fail to mention the denoting of distance, but it is equally difficult to ignore the possibility that the hieroglyphs were measurements or signposts – especially given that *Hermae* were often used in such a way.[16] As desert routes could be easily obscured by sand, it is likely that they also served merely to mark the route.

Desert routes seem to have been well travelled: there is a good deal of evidence for caravans, small groups and individuals plying them on a regular basis.[17] The long distances involved, and the highly arid terrain, meant that camels were the main animals used for travel in the desert, although donkeys are extremely hardy, and often appear travelling and transporting in these areas. Carts and wagons seem to have been used infrequently.

These travellers have left their mark, occasionally in the textual record, but more commonly in the form of graffiti, which adorns many of the rocks and walls of desert locations. Many of these short inscriptions are collected in modern works, and have been discussed in detail.[18] Many are dedications to gods, and most commonly to Pan; but it is probable that their main function was recording one's name for posterity – an ancient 'Kilroy was here'. Some are more interesting than others, especially those which call Pan the 'giver of rock and good routes'.[19] The importance of good routes in the desert was of obvious importance, and the invocation of the god Pan is interesting in its modern etymology, as it has given rise to the French *panique*, which in its more literal sense describes the feeling of loneliness and desolation in desert environments.

In the desert, stations and garrison posts provided the necessary resting points along routes. In the Nile valley, such services could be sought in the cities at inns and at state-run *mansiones*. The latter made up staging points for the state post and for travellers engaged on state business, and likely replaced the *ad hoc* requisition of lodgings so resented by local populations.[20] Two recently published documents dating to the fourth century preserve extensive sections of accounts of the *mansiones* connected with the *cursus publicus* at the village of Tacona in the Oxyrhynchite *nome* and the metropolis of Oxyrhynchus itself.[21] The accounts list the names and number of persons staying at each *mansio*, the numbers of animals accompanying them, and the amount of rations issued (see Figure 7.2).

Entries made in the account follow a standard format: the identity of the travelling party; the time spent in each *mansio* (in all cases one night in Tacona, two in Oxyrhynchus) and the date; a total of days spent in each *mansio*; and finally the number of rations for men (*annonae*) and fodder for animals (*capita*).[22] The normal daily rations for men seem to have been one-sixth of a *modius* of bread, one *sextarius* of wine and half a *litra* of meat; and for animals, half a *modius* of barley and twenty *litrai* of chaff. A careful record of accounts, and a stipulated ration scale, meant that the administrators of the *mansio* could plan their stocks. It is the ration details provided which enable us to calculate the number of persons staying in the mansio on the days for which we have a record:

Figure 7.2 *Mansio* accounts: *P.Oxy.* LX 4088, col.ii (courtesy of the Egypt Exploration Society)

> To the slaves and animals of Valentinus, *scutarius*, travelling up to the Thebaid, in Tacona for 1 day, Pauni 2, and in the city for 2 days, Pauni 3 and 4, total 3 days: daily, 9 *annonae* and 12 *capita*. Bread, 27 *annonae*; wine, 27 *sextarii*; meat, 13½ *litrai*; barley, 36 *capita* = 18 *modii*; chaff, 36 *capita* = 720 *litrai*.
>
> To the same, travelling down from the Thebaid, in the city for 2 days, Pauni 26 and 27, and in Tacona for 1 day, Pauni 28, total 3 days: daily, 10 {*annonae*} and 1[2] *capita*. Bread, 30 *annonae*; wine, 30 *sextarii*; meat, 15 *litrai*; barley, 36 *capita* = 18 *modii*; chaff, 720 *litrai*.[23]

The selection above records the total *annonae* and *capita* given over a period of three days to the travelling party of Valentinus the *scutarius*, and we see that his party consisted of nine men and twelve animals when travelling to the Thebaid, but that one additional man joined the party for the return journey. On some occasions the groups travelling could be large: the largest group consisted of fifty-two people. The number of animals seems usually to have been similar in size, if not slightly larger – so travelling parties were of a considerable size. There was also the possibility that parties may have overlapped, and so it is that on one occasion more than ninety travellers and their animals stayed in the *mansio* at Tacona.

The impression we have, then, is of an extremely busy concern, catering for large numbers of soldiers and officials travelling on public business – indeed, it must have been difficult for officials to provide for such large numbers on a daily basis. It would be interesting to know if such regular traffic was a feature of official business, or if advanced notice was given. Large amounts of supplies were required, and careful accounting practice was essential. But for whom were these accounts drawn up, and was the identity of the travelling parties recorded in order that record could be kept of who submitted warrants for travel to the officials at the *mansio* in return for their rations? It is likely that the answer lies in two extremely valuable papyri from Panopolis, which date to AD 298 and 300.[24] These preserve a large number of letters forming the correspondence of the procurator of the Lower Thebaid, a senior administrative official.[25] The dynamics of administration are complicated: it certainly appears that ultimate authority lay with the procurator in his region – it was he who could issue warrants for the use of state transport facilities, and it was important that those using them could prove that they possessed such warrants. He scrutinised the monthly reports provided by the *mansio* officials, but most importantly, he appointed individuals to perform what must have been liturgical services from lists of names given to him by the town council of Panopolis, the *nome* metropolis.[26] These liturgists were responsible for receiving and organising supplies for each *mansio*, which could include the receiving of food and fodder, but also the supply of bedding.[27] So the burden of the day-to-day running, and the supply of provisions to *mansiones*, was actually undertaken by private individuals in their capacity as liturgists, and was supervised by state officials. The documentation also served as a record of who used the *mansio*, so that a check on abuses of the facilities could be made.

The obvious question is: to what extent this evidence can apply to other parts of the Roman world? Were the *mansiones* at Tacona and Oxyrhynchus similar in design and administration to those found elsewhere? At this stage in our knowledge it is not possible to answer this with confidence, although on balance it would not seem unreasonable to suggest that the system we see in operation here would be widespread. There is no archaeological evidence for these *mansiones*, a familiar problem in the Nile valley, but it is possible that their design was similar to those stations found in the Eastern Desert.[28] It is also becoming clear that documentary practice, at least in relation to military administration, was similar throughout the empire, and it is likely that this was the case with the *cursus publicus* (see Chapter 5).

There are few references in the papyri to inns providing for civilian travellers.[29] Strabo mentions the notorious lodging houses along the Canobic Canal outside Alexandria:

> Eleusis is a settlement near both Alexandria and Nicopolis, is situated on the Canobic Canal itself, and has lodging places and commanding views for those who wish to engage in revelry, both men and women, and is a beginning, as it were, of the "Canobic" life and the shamelessness there current.[30]

What seems certain is that these inns served travellers going to and from Alexandria, and were certainly for civilians. It is the case that a sixth- or seventh-century inn at the town of Ombos, north of Syene, catered exclusively for civilians and their animals, for a plaque records: 'carried out a cleaning of the public building [hostel] and of the vast amount of dung accumulated over so long a time . . . The whole place was renewed and rebuilt from the ground up for the sheltering of strangers and those with no rights of requisition.'[31]

MEANS OF TRAVEL

Travel by land usually took place either on foot or on the back of donkeys or camels. The latter were commonly used for desert travel, but were not indigenous to Egypt – indeed, their widespread use seems only to have begun in the Roman period.[32] Even then, their use seems to have been more common among local Egyptians than their Graeco-Roman masters. They favoured donkeys as transport for local travel when not walking. Donkey-drivers turn up often in the papyrological record, and it seems that these were professionals. In many cases they delivered goods and letters for individuals to friends or relations: this is frequently attested.[33] These couriers existed alongside the traders and transporters of goods, and individuals who specialised in the hire of animals and wagons for the purposes of transport.[34] Animals could also be requisitioned for use by soldiers, imperial officials on state business, and sometimes illegally for private use. In some cases villages were possibly expected to provide animals for such officials in particular regions or along certain stretches of road, which is probably the case in a short document from the Oxyrhynchite *nome*, dating to the fourth century: 'From the *strategos* to the *komarchs* and *archephodos* of Theresis. Provide two donkeys and one guard for the man who delivers this letter to you, as far as Memphis.'[35]

Horses, apart from those used by the army in Egypt, seem to have been rare. They are not highly attested in the papyri: the reasons for this are probably the expense of rearing such animals and the much higher maintenance costs associated with them. As such, they seem only to have been used by rich landowners or imperial officials. Such individuals may also have used carriages, although there is only one reference for this to my knowledge.[36]

River travel and the types of Nile ship and craft have been discussed often by scholars.[37] There is much less evidence for river travel in Egypt than for that by land. This cannot be taken as evidence for it being less important; it is more a symptom of the random preservation of our evidence. What we do have is instructive. Most documents concern the shipping of grain, although there are a small number of private freight contracts. The most interesting points for our purpose here are the stipulations made on such contracts about how river journeys should be conducted:

> Tryphon shall provide a sufficient number of sailors and complete supplies for the boat. He shall anchor at the safest and designated anchorages at the proper hours, and if he does not deliver the cargo, he shall pay the prescribed penalty.[38]

Little river travel would have been possible at night. Ancient sailors on the Nile did not benefit from the lights of villages and towns by the riverside to guide them. Night navigation was therefore impossible, and the owners of their cargoes were hesitant to allow them to travel in such circumstances. A text from the third century BC is particularly revealing here, as it records arrangements for vessels enduring poor conditions on the river:

> But if any, being driven by a storm, are not able to anchor on the promontory when they come to the harbour and its appurtenances, let them announce to the police the reason and the place in which they have anchored. To those who have reported, the chief of police shall send a guard adequate to protect them while they are moored, so that no violence may be done.[39]

So hazards of navigation and wind were not the only concern; river guards existed to help prevent piracy, which was not restricted to the seas, and, as we shall see below, disturbances in regions of Egypt might also have created problems for travellers. Captains of vessels were also responsible for organising adequate supplies for the journey, and in two documents, we may have preserved a bill of loading for such supplies.[40] Money, clothing and food are listed, along with a variety of utensils.

It is reasonable to assume that boats travelling on the river took passengers, and that it was possible to arrange a passage with a ship's captain already engaged in the transport of cargoes, just as a trader named Athenodorus, living during the reign of Augustus, arranged for his sister.[41] Often, of course, it may have been possible for individuals to be carried on boats belonging to a friend or relative – we saw in a letter quoted above that a family in Oxyrhynchus was prepared to send a boat to collect a woman to bring her to the city for a festival.

Often, however, fares would have to have been paid. We have some evidence for this, but we do not have enough to calculate a rate of fare, because no fare is recorded, the point of departure and destination are not mentioned, there is little to distinguish a passenger fare from a freight charge, or the charges seem arbitrary.[42] Occasionally charges were made for groups of people, and it is certain that these were civilians; it is likely that special arrangements were made, for example, for the passage of soldiers, as repayment would have to be sought from the state.[43] Soldiers often lived far from their homes, and one familiar phenomenon is being unable to meet travelling expenses. Thus it was that a woman named Sentis, possibly the wife of a soldier named Proklos and possibly living in the Fayum, sent him the fare home from his base in the Eastern Desert, and a more financially astute soldier from an unknown location was able to send his wife sixty drachmas to pay for her passage by boat.[44]

We may be able to make something of this last document. Soldiers were well paid in comparison to their civilian counterparts: in the early third century, to which this text dates, legionaries were paid about 1,200 denarii (or 5,760 drachmas) per annum.[45] While some soldiers, like those mentioned above, may have been unable to meet expenses, many would have had savings, and depending on status and length of service, these could be quite substantial. Sixty drachmas, therefore, was not a great deal of money for a soldier. However, it may have been as much as two times the normal monthly income of a labourer.[46] The journey being made must have been long – we have no provenance for the text, but it is likely from internal evidence that the woman lived in the Fayum, and possible that her husband was based in the Thebaid or Alexandria.[47] Similar distances may be implied by a second-century text which concerns the journey of a man named Sarapion from the Fayum to, possibly, Alexandria.[48] Sarapion was unable to return directly, possibly because he had run out of money. Accordingly, his wife Selene sent him 200 drachmas, 54 of which were to pay for his return fare and taxes (which may have been tolls). Travel may then have been expensive in comparison to normal living costs, and the major question, therefore, is who could afford to travel around the country? Soldiers and veterans and their families were well-to-do and could in most cases easily afford to travel, but it may well have been the case that journeys over long distances were beyond the means of the lower socio-economic classes. This would mean a severe limitation on their mobility. Not only was travel expensive, but it also meant time away from their land and livelihood.

It is difficult to come to any conclusions about the relative cost of land and river travel for individuals. More reliable estimates can be made when considering the transport of goods, as in a small number of cases freight rates have been preserved. All we can say about travel is that, for long distances, travel by river would have been quicker, and therefore cheaper. In one third-century document from Philadelphia in the Fayum, a man writes to his wife, telling her that he had arrived safely in Alexandria after four days' travel.[49]

There is always, however, the chance that ship captains, realising this, increased their fares accordingly.

The Nile, while a great highway, was also an obstacle. There were no fording points and bridges would have been impossible to build; so without the provision of ferryboats, crossing the river would have been virtually impossible. Ferries are attested only rarely in the papyrological record, and from these documents it seems that they were tightly controlled by the government. Early in the Roman period, concessions were granted by the state to ferrymen, and at some point a tax was introduced for ferryboats, which, rather than being a tax on a monopoly, seems to have been a licence tax for the operations of a privately owned vessel.[50] Why should this be the case? Was it simply a convenient way to generate income for the state, in the same way that transit tolls of the Eastern Desert routes did, or was there some attempt to restrict movement, as crossing the river would invariably mean crossing *nome* boundaries? Most likely it was simply another tax, in a country were such taxes were common.

REASONS FOR TRAVEL

Private letters from Egypt record information which is of great interest to social historians. The main restriction of the evidence, however, is that it is largely anecdotal and often lacks any context. Most letters simply served to advise friends and relatives of the writer's health, or perhaps were requests to send items of clothing or food. In some, however, we are given more information concerning reasons why people travelled through Egypt, and where the writers are not explicit, we can often infer something of their business. The purposes of travel fall into two main categories: private business (which may include family matters or private trade); and public business (journeys necessary for performing state business or attending court hearings).

Perhaps what is most striking about private letters from the Roman period is the humanity and warmth expressed between correspondents. These letters, more than any other form of evidence, bring us close to the everyday concerns of ancient people, and demonstrate that their lives were punctuated with the same feelings and cares as ours today. There are large numbers of examples, but we should isolate a few. Family events come high on our list of private reasons to travel. In the early second century, a soldier based in the Eastern Desert wrote to his wife in the Fayum, who was in the final stages of her pregnancy, to tell her that he would do his best to organise leave in order to visit her. He planned to use the provisions boat which presumably brought grain and other supplies from the Fayum to the Thebaid.[51] The availability of boats or other forms of transport was clearly an important consideration, and in a similar letter from the fifth century we read of a brother who intended to visit his sister, but only if he could find a boat.[52] Private letters also contain many requests from relatives that other members of their family should visit, and for many reasons, which may include birthdays or festivals, or perhaps simply company. In an important group of letters exchanged between a man named Paniskos and his wife Ploutogenia, a particularly interesting letter asks her to visit him in Koptos:

> Paniskos, to my wife, Ploutogenia, mother of my daughter, very many greetings. First of all I pray daily for your good health in the presence of all the gods. I would have you know then, sister, that we have been staying in Koptos near your sister and her children, so you may not be grieved about coming to Koptos; for your kinsfolk are here. And just as you desire above all to greet her with many greetings, so she prays daily to the gods desiring to greet you along with your mother. So when you have received this letter of mine make your preparations in order that you may come at once if I send for you. And when you come, bring ten shearings of wool, six jars of olives, four jars of liquid honey, and my shield, the new one only, and my helmet. Bring also my lances. Bring also the fittings of the tent. If you find an opportunity, come here with good men. Let Nonnos come with you. Bring all our clothes when you come. When you come, bring your gold ornaments, but do not wear them on the boat.
>
> [verso] I salute my lady daughter Heliodora. Hermeias salutes you.
>
> [address] Deliver to my wife and my daughter, from Paniskos her father.[53]

Paniskos writes several times to his wife, in each letter asking her to bring particular items. The reasons for Paniskos' protracted stay are unclear. That he asks for lances and helmets to be brought may indicate that he was a soldier, and this could perhaps be linked to disturbances in the Thebaid at this time – but it is more probable that he was involved in the revolt of Achilleus, fighting on behalf of the rebels.[54] Although the fact that the letter arrived in the Fayum shows that travel was possible, the remainder of the archive makes it clear that Ploutogenia herself never seems to have journeyed to Koptos. Her husband implies that she is to come only if she can find good and trustworthy men, perhaps because it was dangerous; although it is entirely possible that Paniskos is asking his wife to recruit supporters for the rebels. It is notable that he makes a particular request of his wife that she does not openly display her jewellery on the boat.

Festivals and holidays were important times for families to be together. We have seen already the letter asking a woman to come to Oxyrhynchus for the 'birthday' of the god. Public holidays were certainly thought worthy of travelling some distance for, and demands for entertainers to travel were frequent. An interesting request from Oxyrhynchus is typical:

> The Aurelii Agathus, *gymnasiarch* and *prytanis* in office, Hermanobammon, *exegetes*, Didymus, chief-priest, and Coprias, *kosmetes* of the city of Arsinoe, to the Aurelii Euripas, actor, and Sarapas, Homeric reciter, greeting. Come at once, in accordance with your custom of taking part in the holiday, in order to celebrate with us our traditional festival on the birthday of Kronus the most great god.[55]

Family birthdays, especially the first birthdays of children, were times of celebration, and invitations were extended to family members to dine.[56] Less joyful family occasions similarly demanded travel. A number of letters preserve details of journeys made to attend family funerals or ill relatives.[57]

Travel on business was clearly important, and documents provide much evidence for this. Customs-house receipts, for example, show that donkey- and camel-drivers regularly plied the routes of the Western Desert, and those of the Eastern Desert were similarly busy. Indeed, there is good reason to think that transport was performed by a substantial number of professionals. This form of travel lies outside our interests here, as we are more concerned with travel rather than the economics of transport, but we should consider briefly more ambitious journeys made in the interests of business. One letter is particularly well known in this respect, and any modern reader will recognise the concerns of the writer:

> Antonius to his dearest Dionysius, greeting. I felt it incumbent upon me, as soon as I set foot in Italy, to let you know that I and my family are in good health. We had a slow passage, but not a bad one, and my poor self experienced no discomfort such as usually occurs, particularly on one's first voyage. [second hand] . . .
>
> [Third hand] Written in Puteoli, Pauni 4.
>
> Addressed: To Dionysius, overseer, from Antonius, son of Ptolemaeus.[58]

No mention is made of the purpose of Antonius' journey, but the fact that he writes to Dionysius the overseer may suggest business interests. The nature of private letters unfortunately means that we will never know. But is clear that Antonius and his family were making their first sea journey, and that they had been worried about seasickness. A journey to Rome was not undertaken lightly, as it meant not only expense, but also certain legal requirements, as we shall see. Often journeys to Rome were made by recruits to the army, which is evident from a group of letters written from a son to his mother, a resident of Karanis in the Fayum:

> Apollonarios to Taesion his mother, many greetings. First of all, be in good health for me, as I am making obeisance before all the gods. From Cyrene I encountered a man travelling in your direction and I felt it necessary to tell you about my well-being. You, too, tell me soon about

your well-being and that of my brothers. Even now from Portus I am writing to inform you, for I have not yet gone up to Rome and been assigned. When I do get my assignment and know into which unit I am going, I will inform you immediately. You, too, do not hesitate to write to me concerning your safety and that of my brothers. If you do not find a man travelling towards me, write to Sokrates and he sends it on to me himself. I greet my brothers many times and Apollonarios and his children, and Kalalas and his children, and all those who love you. Asklepiades greets you. Farewell in good health. I arrived at Portus on Pachon 25.

{Second hand} Know that I was assigned to Misenum, for I learned it later.

{Address on back} Deliver to Karanis, to Taesion, from Apollonarios her son.[59]

Apart from this letter's importance to our understanding of the assignment of recruits to military units (although it probably throws up more questions than it answers), it provides valuable information about how private letters were delivered. Apollonarios met a man who was travelling to Egypt, and perhaps Karanis, and was able to give this letter to him for delivery. It is likely that this private arrangement was the most common way for letters to be delivered, although, for official post, a postal service did exist.[60]

But how, if the individual carrying a letter was unfamiliar with his destination, did he know where to deliver it? Word of mouth, of course, was perhaps the usual way – it was easy to ask for directions. But in one unique document from Oxyrhynchus, we have preserved a series of directions for the delivery of post to a man named Rufus, which incidentally provides useful information on the urban topography of a city for which no good archaeological evidence remains:

Directions for letters to Rufus.
From the Moon Gate walk as of towards the granaries and, if you will, turn left at the first street behind the Thermae, where (there is) a . . . , and go westwards. Go down the steps and up . . . and turn right after the precinct of the (temple) on the right there is a seven-storey house and on the top of the gatehouse (a statue of Fortune?) and opposite a basket-weaving shop. Enquire there or from the concierge and you will be informed. And shout yourself . . . will answer you.[61]

Before moving on to consider travel on public business, we should discuss the valuable evidence for travellers provided by a very well-known inscription from Koptos.[62] Often used as evidence for commerce, the Koptos Tariff inscription is perhaps more to do with travel by civilians than anything else. It records charges made by the state for the use of routes from Koptos to the Red

Figure 7.3 Koptos Tariff inscription. Photograph J.G. Milne 1924

Sea ports, and shows what kinds of people used them (Figure 7.3). As we would expect, high on the list come sailors and ships' captains; animal-drivers had to pay tolls for their animals and wagons. But much more interesting are charges made of the wives of soldiers and sailors, and prostitutes; charges which were much higher than for any other road-users. Women, it seems, did travel into the desert, and *ostraca* from Mons Claudianus dating to the early second century show that they used desert routes there.[63] This is interesting not only because women were visiting their 'partners', but also because it was only from the reign Septimius Severus that soldiers were legally permitted to marry. It seems likely that the state tolerated this natural state of affairs, but could also profit.

TRAVEL ON PUBLIC BUSINESS

Travel on public business was an important part of life in Roman Egypt, and has thus received more attention from scholars than other forms of travel. The performance of liturgies, or compulsory public services, often entailed individuals travelling away from home. A few examples here will suffice. A considerable number of documents relate to the supply of grain and animals to the army.[64] It was common for convoys of supplies or of animals destined for use either with the army or for the transport of state grain to travel between *nomes*. One incident which took place in the second century not only suggests the onerous nature of liturgies and of the transport of grain, but also illustrates the limitations of the state's control over individuals. The *strategos* of the Polemon and Themistos divisions of the Arsinoite *nome* makes a request of the *strategos* of the Oxyrhynchite:

> If you were present when the most illustrious prefect threatened the *strategoi* with regard to the transport of the corn, to send as many animals as possible from the other *nomes* to work in the Arsinoite, they had proceeded against the herdsmen who had presumptuously run away after the order of the prefect . . . although there are only 411 donkeys here from your *nome*, most of them have run away, so that up to the present only 156 remain . . . send an equal number with herdsmen of standing who can stay, in order that, while the river is still navigable, the transportation may be carried out, because the water is already imperceptibly rising and the need is urgent that the corn be very quickly brought down.[65]

Another example is of interest here. Towards the end of the second century, a veteran cavalryman named Dionysius Amyntinus had been assigned as a liturgist to the escort of blankets destined for the legion in Alexandria. Once he had arrived he was detained for over forty days, and his worry is evident. He petitions the *praefectus castrorum* to take the blankets 'so that I too, now that the time for sowing is at hand, may return to my home, in order that I may receive relief'.[66] The time spent away from one's land was clearly a matter of concern, as has already been mentioned.

Delays were a common cause for concern. In the first century, a man whose name is lost wrote to his father begging assistance, as a friend's wife was pregnant and coming to the end of her term, but he had been delayed during court hearings concerning the receiving of legacies.[67] In another document, a man and his wife write to a friend saying that they cannot attend the wedding of her son, as they were detained by illness and the prefect's *conventus*, his annual tour of Egypt during which legal cases and petitions were heard.[68]

PROBLEMS AND RESTRICTIONS EXPERIENCED DURING TRAVEL

Travel in Egypt was not without its difficulties, which could arise from natural phenomena or more sinister dangers.[69] Although in many ways tragic, some of the events recorded are certainly comic.

Sometimes it was simply not possible to travel. The most common reason was the Nile flood. During this time, according to Herodotus, the Nile valley resembled a sea, with islands dotted throughout. This finds visual expression in the famous Palestrina mosaic from Italy (Figure 7.4).[70] The flood made river transport very difficult, and indeed, at least in pre-Roman times, there may have been a royal taboo on sailing during the inundation.[71] But it also had a profound effect upon travel along roads in the Nile valley. The extent of damage caused by the flood is evident from the fact that all land measurements and divisions had to be re-measured each year, and all irrigation works had to receive substantial repairs. This is clearly the main reason why roads in Egypt were merely tracks – the maintenance of anything more would have been impossible. A number of private letters refer to problems with roads being impassable.[72]

There were other reasons why roads could be impassable. In a letter exchanged between two business associates in the late third century, one records that he had been 'advised by the most notable Ammonion to send for a ferry-boat on account of the uncertainty of the road'.[73] This may have been due to the flood or similar natural causes, but in the late third century, disturbances in Egypt are well documented; indeed, we saw above that Paniskos and his wife Ploutogenia may have experienced problems with their travel plans due to fighting. In another text from the same period, a man writes to his wife reassuring her that there is no fighting in the region lying between them.[74] Later in the same letter, however, he is more guarded: 'If there is fighting going on in between us, write to me and events will decide. Otherwise, do you come alone – always provided there is no fighting.' But such revolts and disturbances were not the only hazards which travellers had to endure. Banditry and similar violent crime were common features of life on the road.[75] Some time in the third century, Petesouchos, son of Pasis, was travelling from the village of Tebctnou to Corphotoi in order to visit his sister, when he was attacked by bandits who stole his clothing and money.[76] Drunks, of course, were another problem, and could often be encountered on the road, and this frequently resulted in insult, injury or theft.[77] The only option in the first instance was to seek the help of guards, but they were not always sympathetic: in one document a traveller complains that when his two donkeys were stolen, to add insult to injury, the guards of the village of Bacchias smashed his water jars and threw him in jail for three days, probably because they were acting in collusion with the bandits.[78] With such occurrences possible, we can understand the concern of Paniskos that his wife should keep her jewellery out of sight on her planned journey to see him in Koptos. There was a carefully structured system of

Figure 7.4 Nile Palestrina mosaic detail. Photograph R. Laurence

watchtowers throughout the country, garrisoned by soldiers in the Eastern Desert, and by liturgists in other parts of the country.[79] These served not just to protect travellers, but to guard trade routes.

There were other hazards, and like today, roads were not safe. So it is that we have a record of an early road-traffic accident on the streets of Oxyrhynchus in AD 59:

> To Tiberius Claudius Ammonius, *strategos*, from Lucius Panisius, veteran. I rode up, on donkey-back, from the fields to the city of Oxyrhynchus, on the first of the present month Pharmouthi, in the 5th year of Nero Claudius Caesar Germanicus. A slave of Sarapion, son of Dorcon, was accompanying donkeys loaded with rocks which collided with me next to . . . One of the donkeys kicked me violently on the lower right leg. I have since been confined to bed, and my life is in danger. And as the slave himself saw his bad luck, he has fled, and I have retained the animal and it is held at my house. After that Sarapion demanded that I return his donkey. That is why I have made this demand.[80]

Such injuries, despite the rather emotive tones of this petition, could be serious, and it is natural that one should seek redress. Sometimes travellers took the law into their own hands, and it is clear to see why in the following case:

> To King Ptolemy, greeting. I have been unjustly treated by Psenobastis, who lives in Pysa in the division of Herakleides. For in the 4th year, on Phamenoth 21, I went to Pysa on private business. An Egyptian woman, whose name is said to be Psenobastis, having leaned from the upper storey emptied a chamber pot into the street so that I was soaked. And when I became angered and reproved her, Psenobastis pulled at my cloak with her right hand and uncovered my chest, and spat in my face. I have witnesses to prove this and I have been the victim of an unjust attack.[81]

Such problems, whether they be caused by flood, disturbance or some more mundane cause, could often mean that people were unwilling to travel, or that it was impossible to persuade professional transporters to accompany travellers. It is no coincidence, therefore, that a particularly interesting letter from the third century (again) details problems which a man encountered on his journey north to Oxyrhynchus:

> I have not been able to find a means of coming to you, since the camel-drivers were not willing to travel to the Oxyrhynchite *nome*. Not

> only that, I came up to Antinoopolis and was unable to find a boat and I was unable to do so. So now I have made plans for my baggage to be sent to Antinoopolis and I will stay there until I find a boat and then I will sail.[82]

Unfortunately we cannot be precise about the date of this letter, but it seems reasonable to assume that the camel-drivers were unwilling to travel because of rebellion or disturbances – especially if the man was travelling from the Thebaid, which was a hotbed of revolt throughout the Roman period.

Other problems which travellers encountered were restrictions placed on movement by the state. This is an enormous historical problem, and has been the subject of much research.[83] A great deal of this work, however, has been concerned with establishing the mobility of individuals in terms of their area of employment in relation to their *origo* or *idia*. It has focused less on the evidence for travel contained in private letters, and has largely avoided evidence for recreational or occasional travel. In some respects a result of this is that our view of life in the Roman world, rather like that of the medieval world, is that of a society in which travel, and certainly recreational travel, was uncommon.[84] Indeed, the lot of the lower socio-economic levels in respect of mobility has been described thus: 'another feature we associate with traditional peasantry is geographical immobility'.[85] But the evidence from Egypt which we have considered so far shows the contrary – that there was a fair degree of mobility in Roman Egypt. People did travel, and they were often relatively humble. It is not fully clear, given certain limitations of our evidence, to determine just how far down the social scale travellers could be found, but I suspect that even the relatively poor could find some time and opportunity to travel at least short distances. And it is certainly the case that recreational travel was not extraordinary. In one papyrus from Oxyrhynchus, the writer clearly has a recreational journey in mind when he writes to his friend: 'I think that the idea of getting away seems reasonable to you too, since the weather is already getting hot.'[86] And it is certainly the case when a man named Herodes asked permission of his employer if he could take two days' holiday to visit Alexandria, perhaps for family reasons.[87]

But although society was mobile, there were restrictions. The most obvious one is that, famously, Roman senators and many equites were forbidden to enter Egypt after its annexation by Augustus.[88] How far this measure was enforceable in reality is hard to measure, but it is clear that this restriction did not extend to others, which can be seen from a letter of Pliny the Younger concerning his freedman Zosimus, whom he had sent to Egypt to recover from illness.[89] Perhaps less easily explained are the restrictions placed on individuals leaving Egypt:

> To Valerius Firmus, prefect of Egypt, from Aurelia Maeciana of Side (in Pamphylia). I wish, my lord, to sail out by way of Pharos; I beg you

> to write to the procurator of Pharos to allow me to leave according to the usual practice. Pachon 1. Farewell. [There follows remains of a permit written in Latin from the prefect].[90]

Other evidence of similar regulations shows that, rather than being mere restrictions, passes were allocated according to the nationality of those leaving Alexandria, and thus we have, in many respects, an early precursor to a modern passport. Our passports give us the right 'to pass freely without let or hindrance', but they also act as identification and proof of national status. These charges generated income for the state, as visas do today, and it is likely that they were similar in character to the charges already seen on the Koptos Tariff inscription. But they also served to regulate travel, and we can connect them to similar passes found recently at Mons Claudianus, which allow travellers to use the routes near the quarries in the Eastern Desert.[91] Rather than providing evidence for restrictions on travel, they do the opposite, and show that travel was an accepted and common aspect of life, and one from which the state could generate income through the charging of transit tolls.

PATTERNS OF TRAVEL

The nature of papyrological evidence allows for some quantification. Alston has used evidence from Oxyrhynchus to establish a simple communication network for this city.[92] His evidence, however, is not without problems. Only 201 documents preserve enough information for us to know points of destination, and these documents are spread over a period of six centuries. It is not possible, therefore, to establish a real estimate of the volume of traffic, but only to give a flavour of what patterns of transport may have existed. We must also remember that the purpose of these journeys varied; most are made in connection with trade, and therefore are of economic importance and do not properly reflect how mobile the culture was beyond trade.

More than half of all attested journeys were made to the north, and of these most, as we would expect, were to Alexandria. This pattern remains the same if the documents are split into groups according to century. Some 45 per cent of these journeys are made within a radius of 90 km, while only 17 per cent are made to destinations about 260 km north or south. So most journeys are of reasonably short duration. Only 13 per cent of journeys are made to destinations further away or outside Egypt.

Similar patterns can be seen at Karanis, another village for which we have a substantial body of evidence. There are clear economic links with neighbouring villages and beyond, and from tax rolls preserved from another Fayum village, Philadelphia, we know that between 4 and 13 per cent of the adult male population at any one time was made up of resident aliens from other villages. At Karanis, most (60–75 per cent) communication with other villages took place within the Fayum, regular links were established with Memphis and the Herakleopolite *nome*, but these too lay close by. Between 10 and 20 per cent of communication was with Alexandria, and 6–10 per cent was external (we have already seen examples from this village of travel to Rome). Another Fayum village, Soknopaiou Nesos, which lay on the boundaries of the Arsinoite nome beside an important desert route, had more regular links with other areas.[93] Here we have regular contact documented with the Western Oases, Memphis, the *nomes* lying to the north and, in two cases, with Alexandria. But nearly all of this communication is recorded on customs-house receipts and is therefore of economic importance only. But these two villages are peculiar in their own ways, and this limits the applicability of their evidence. Karanis, or at least the documents which have been preserved, indicate a high proportion of Roman citizens, and of these a large number of veteran soldiers.[94] This may explain travel outside Egypt, and certainly journeys made to Rome in order to join the army or navy. At Soknopaiou Nesos the opposite is the case.[95] Nearly all of the population were Egyptian, and would therefore not have had the same links outside their village as veterans may have had, but the villagers seem to have based their livelihoods on trade and transport, and our evidence would reflect this. These caveats aside, however, the sum of our evidence shows a population which was fairly mobile for different reasons. Most communication seems to have taken place within, say, a 100 km radius, but a reasonable amount beyond. In a pre-industrial society, this is what we would expect. The importance of Egyptian evidence, however, lies in the fact that it shows us that travel and mobility were not solely the preserves of the rich, but that, for many reasons, all but perhaps the very poorest could travel if need be.

TRAVEL ITINERARIES

Only three travel itineraries are preserved on papyrus.[96] The most important is preserved as part of the so-called archive of Theophanes, which is made up of public papers, private letters, itineraries of journeys, accounts and various memoranda, which all seem to have come from the city of Hermopolis in Middle Egypt. Also important is an itinerary preserving details of a journey made from Alexandria to, most likely, Oxyrhynchus. Finally, a private letter mentions an itinerary made through Asia Minor to Egypt, but no details are preserved.

The Oxyrhynchus itinerary is a fragmentary text of the first century AD. The travelling party appears to have been in no hurry, as they take time off on at least five days of their journey in order to bathe. Their route takes them from Alexandria into Egypt (Alexandria being thought separate from the rest of the Egyptian *chora*) at Nicopolis, from there they travel by canal and river through Schedia to Hermoupolis Minor, past Babylon and Memphis to Aphrodito, along the Fayum Canal to Venne, to Ptolemais Hormou, where they joined the Bahr Yusuf, which they followed past Kaine and Tacona to Oxyrhynchus. That they travelled by boat is certainly suggested by the fact that they stop at Memphis on the west bank of the Nile and Babylon and Aphrodito on the east bank. The journey took nine days from Nicopolis to Memphis, whereas by land the same journey took five, according to the Antonine Itineraries. The relaxed pace of travel suggests that the travellers were not engaged in business. When compared to the Antonine Itineraries and the Peutinger Table, different cities and stopping points appear, but Memphis appears in all three, while Venne appears in the Peutinger Table, and is unattested anywhere else apart from our document.[97]

More instructive is the itinerary preserved as part of the archive of Theophanes, who may have held the position of adviser or legal advocate to a high official, and perhaps was part of the staff of the prefect of Egypt. Two letters of introduction written for him and addressed to provincial governors in Syria indicate that he must have been a man of some standing. His itinerary took him from Antinoopolis to Antioch in Syria, and his letters may have introduced him to the *praefecto praetorio per Orientem*, a superior official, and Achilleus, the governor of Syria. There is no clear evidence for the purpose of his journey, but the date of the archive generally falls within the second and third decades of the fourth century, and the prices of commodities listed on his expense accounts, to which we shall turn in a moment, suggest a date in the early 320s.

We have travel accounts for the outward journey and an itinerary, which is incomplete, but we have been lucky in that a draft copy of it was preserved with the accounts.

There is much of interest here. The distances noted by Theophanes often tally and are sometimes identical with distances preserved between the relevant stations on the Peutinger Table and in the Antonine Itineraries. The distances between stations were generally longer than we may find elsewhere in the Roman empire, and once Theophanes left Egypt, he appears to have covered more miles in one day than he generally travelled when in the Nile valley. We could attribute this perhaps to fewer distractions on the barren coast of Palestine, and an increasing urgency in his journey, perhaps because he had travelled slowly in Egypt. Theophanes' journey through Egypt was rather convoluted. It is likely that the first stage from Middle Egypt to Alexandria (for which few details remain) was by river. From Alexandria he re-enters Egypt (Alexandria being separate from Egypt) at Niciu, from where he follows a route not recorded elsewhere until he joins the main route from Babylon to Pelusium at the city of Herakleopolis Parva. From there his route follows the road well known from other sources. His journey began perhaps in early March, and averaging about 32 miles per day, he reached Babylon on 19 March, and eventually Antioch in Syria on 2 May, where he spent two months before beginning his return journey.

Table 7.1 Travel in Roman Egypt: distances according to various sources (after P. Ryl. IV 627 Commentary)

	Location	P. Ryl. *IV 627*	*Peutinger Table*	*Antonine Itinerariers*	*Bordeaux Itinerary*
1	Niciu to Athribis	–	–	–	–
2	Athribis to Leontoplis	20	–	–	–
3	Leontopolis to Thmouis	24	–	–	–
4	Thmouis to Tanis	24	–	22	–
5	Tanis to Herakleopolis Parva	21	–	22	–
6	Herakleopolis to Pelusium	24	[22]	22	–
7	Pelusium to Geras	10	8	–	–
	Geras to Pentashcoenum[1]	12	–	–	–
8	Pentaschoenum to Casium[2]	16	–	20	–
9	Casium to Ostracine	26	–	26	–
10	Ostracine to Rhinocolura	24	23	24	–
11	Rhinocolura to Boutaphium	[15]	–	–	–
	Boutaphium to Raphia[3]	[23]	–	–	–
12	Raphia to Gaza	24	22	22	–
	Gaza to Ascalon	15	15	16	–
13	Ascalon to Jamnia	20	22	20	–
	Jamnia to Lydda	12	12	12	–
14	Lydda to Antipatris	11	12	–	10
	Antipatris to Bettara[4]	17	16	–	10
	Bettara to Caesarea	16	–	18	16
15	Caesarea to Ptolemais	44	28	44	[38–9]
16	Ptolemais to Tyre	45	32	32	32
17	Tyre to Sarepta	16	–	–	[12]
	Sarepta to Sidon[5]	20	–	–	[13]
18	Sidon to Beirut	34	29	30	28
19	Beirut to Bilbus	24	–	24	12
20	Bilbus to Tripolis	36	32	32	32
21	Tripolis to Arcae (Orthosia)	16	12	18	16
22	Arcae to Antaradus	30	30	32	32
23	Antaradus to Balanea	24	24	24	26
24	Balanea to Gabala	14	16	24	13
	Gabala to Laodicea	20	11	18	14
25	Laodicea to Hydata	50	–	48	–
	Hydata to Daphnae	10	–	11	11
	Daphnae to Antioch	4	–	5	5

Notes:

[1] The distance from Pentaschoenum to Pelusium is given as 20 miles in the Antonine Itineraries.

[2] The distance from Pelusium to Casium is 40 miles in the Antonine Itineraries, and 31 miles on the Peutinger Table. The distance from Geras t160o Casium on the Peutinger Table is 23 miles.

[3] The distance from Rhinocolura to Raphia is given as 34 miles on the Peutinger Table, and 22 miles on the Antonine Itineraries.

[4] The Antonine Itineraries give the distance from Bettara to Lydda as 28 miles.

[5] In both the Peutinger Table and Antonine Itineraries the distance between Tyre and Sidon is given as 24 miles.

The accounts preserve details of his expenditure, but it is difficult to establish the size of his entourage (however, we should remember here the size of travelling parties detailed in the *mansio* accounts discussed above). It is clear that he was accompanied by his slaves, for he spends some 900 drachmas per day on their provisions. He also retained the services of two other officials, a steward, a number of clerks and perhaps an official in charge of finances. He himself spent between 2,000 and 3,000 drachmas per day on a variety of items, mainly luxuries. His daily spending on food and wine made up most of his expenditure, but much also was spent on bathing, soaps and papyrus. He also purchased clothes – the editor refers to him as 'something of a dandy'. Presumably his baggage train was of a considerable size – we know from other documents that preparations for journeys entailed the storing of money, clothing, and food for carriage on ship. There are, however, no details preserved in the accounts for how Theophanes organised the transport of his belongings, or indeed of how he travelled. The editor suggests that this was simply because he was using the imperial postal service, the *cursus publicus*, and that transport would thus be supplied by the state. We know that animals were requisitioned for work on the postal service, but the extent of our knowledge about requisition is not sufficient for us to be certain of the arrangements (cf. Chapter 5). What is clear from the documents discussed above is that many officials using the *mansiones* of the *cursus publicus* were travelling using their own animals. Perhaps senior officials were not expected to do this, and we are reminded here of the stipulations on the entitlement of travelling officials detailed on the edict of Libuscidianus in Pisidia.

Several questions arise: why were itineraries recorded, and where did the necessary information come from? The Oxyrhynchus itinerary is much less detailed than that of Theophanes. No distances are recorded, but the number of days travelling does seem to be. It could merely be some form of diary or record, but ultimately the document is too lacunose to draw any conclusions. Theophanes' itinerary is certainly different in character and purpose. The itinerary preserves the names of the stopping points and the distances between them. There are no directions for travel; the itinerary is no map. If the itinerary had been preserved wholly as a separate document, its purpose may have been more difficult to divine. The fact, however, that a draft version is made alongside Theophanes' travel accounts may indicate the purpose in keeping such records of route, and may offer an explanation of one purpose which itineraries may have served more broadly. The two must be connected: Theophanes was allocated funds and most likely a warrant to make his journey and to break it at specified stops using the facilities of the *cursus publicus*. He therefore had to make accurate accounts of both his travel and stops, and the amount spent on himself and his entourage. This means that his superiors, or at least their staff, would have had records against which to check the details of Theophanes' account. It is likely, therefore, that records of state routes were kept in most state offices throughout the empire to assist officials planning official journeys.

Such itineraries provided the information which has been handed down to us in the form of the Antonine Itineraries and Peutinger Table. The nature of these itineraries is discussed elsewhere in this volume (see Chapters 2 and 3), but it is clear from papyri that they were much more than just maps. Although they did not contain directions, they were easy to use to help one get from A to B, but more importantly, they were an important source of information for the state, which could be used to plan journeys, to estimate the cost of journeys made by state officials, and to check that they were not abusing the system by spending too much on their subsistence.

CONCLUSION

This chapter has been wide-ranging, and must serve only as a preliminary study of the evidence for travel preserved on papyrus. The picture which emerges is of a society which was much more mobile than we might think, and certainly more dynamic than the received impression of feudal or early medieval societies. The purposes of travel mentioned on private letters were at once diverse and recognisable to us today – the problems and hazards similarly so. Papyrological evidence provides an insight into who the travelling public were, something which is not possible with other forms of evidence. But papyri are also of value in that we can compare the evidence they preserve with what we know of travel and geography in other parts of the Roman empire, and in the case of *mansiones*, itineraries and the *cursus publicus*, the evidence so provided is of immense value.

NOTES

1 I would like to thank Tom Harrison, Jon Coulston and the anonymous referees for comments on this paper. Any views expressed are solely those of the author.
2 Strabo XVII 1.16.
3 *P. Mich.* VIII 496 (second century).
4 Bagnall (1985); Adams (forthcoming).
5 *P. Oxy.* I 112 (third or fourth century).
6 Suetonius, *Aug.* 49.3; Herodotus V 52–3 on Persia; *P. Hib.* I 110 = *W. Chr.* 435 = *Sel. Pap.* II 397 (third century BC).
7 Sidebotham (1986); Adams (forthcoming).
8 Sidebotham (1986) 58–62.
9 *IGRR* 1142 = *OGIS* 701; Kennedy (1984); Sidebotham and Zitterkopf (1997).
10 Strabo 17.1.45; Pliny, *NH* 6.26.103.
11 Strabo 17.1.13.
12 See Drower (1985) 123.
13 Sidebotham and Zitterkopf (1997).

14 Strabo 17.1.50.
15 Pocock, *Travels in Egypt*, in Pinkerton (1818–24) XV, 265.
16 See Pausanias III 1, with Romaios (1904–5). Pausanias crossed from the Argolid into Lakonia, passing *Hermai* on Mount Parnon, where, in the Roman period, the boundaries of Arkadia, Lakonia and the Argolid met.
17 For example, *P. Vindob.* Gr. 40822 (early second century); the so-called archive of Nicanor, *O. Bodl.* I 220–304, *O. Brüss. Berl.* 7, O. *Bodl.* II 1969–71 (all first century); *OGIS* 676 (AD 90).
18 Foertmeyer (1989) 11–14.
19 *I. Kosseir* 141, 149, 158, 170.
20 See Mitchell (1976) for a discussion of relevant documents.
21 *P. Oxy.* LX 4087 and 4088 (fourth century), see Fig. 7.2.
22 Rea *et al.* (1985).
23 *P. Oxy.* LX 4088 (*c.* AD 347–50) ll. 37–44; *P. Oxy.* LX 4087 ll. 168–70.
24 *P. Beatty Panop.* 1 and 2 (AD 298 and 300).
25 On administration, see Bagnall (1993) 62–7.
26 Bowman (1971) 69–83.
27 *P. Beatty Panop.* 1 ll. 223–5, 263–4.
28 Sidebotham (1986); Maxfield (1996); Peacock and Maxfield (1997).
29 For a general discussion of inns and restaurants in the ancient world, see Casson (1974) 197–218. Papyri: *SB* IV 7475; *P. Iand.* 17; *PSI* I 175.
30 Strabo 170.1.16.
31 *SB* IV 7475.
32 See Adams (forthcoming).
33 For example, *P. Oxy.* XLVII 3357 (late first century).
34 See Adams (forthcoming).
35 Mitchell (1976); Adams (forthcoming); *P. Oxy.* IX 1193 (fourth century).
36 Strabo 17.1.50.
37 Casson (1971).
38 *P. Ross. Georg.* II 18 (AD 140).
39 *P. Hib.* II 198 (third century BC), with Bagnall (1969).
40 *P. Coll. Youtie* II 84 (fourth century); *P. Oxy.* XVI 1923 (late fifth or early sixth century).
41 *BGU* XVI 2604 (7 BC).
42 *P. Dub.* 16 introduction; Meyer-Termeer (1978) 65 n.3, 78 n.5.
43 *P. Lille* I 25 (third century BC), or for individuals *P. Brem.* 48 (AD 118).
44 *O. Flor.* 17 (second century); *P. Dub.* 16 (second or third century).
45 Alston (1994) and (1995) 103–8; soldiers were paid more after the Severan reforms, but the rate of pay is unkown.
46 See Johnson (1936) 306–10 for wage levels.
47 Soldiers were often based long distances from home (Bagnall (1976) 30).
48 *P. Amh.* II 131 (second century).
49 *BGU* VII 1680 = *Sel. Pap.* I 134 (third century).
50 *BGU* IV 1189 (*c.* 15 BC); *P. Oxy.* IV 732 (AD 150).
51 *O. Flor.* 14 (early second century); on leave, see Speidel (1985).
52 *P. Amh.* II 144 (fifth century).
53 *P. Mich.* II 214 (AD 296).

54 See Thomas (1976) 266–7.
55 Lewis (1983) 102–6; *P. Oxy*. VII 1025 (late third century); see also *P. Tebt*. II 412 = *Sel. Pap*. I 128 (late second century), travel arrangements for a new year festival; *P. Oxy*. XIV 1666, for a soldier's plans to visit his family for the festival of Amesysia. Consider also letters concerning the hire of dancing girls and musicians, for example, *P. Grenf*. II 67 (AD 237).
56 For example, *P. Oxy*. XXXVI 2791 (second century).
57 For example, *P. Fouad* I 75 (AD 64).
58 *P. Oxy*. XVIII 2191 (second century). See also *BGU* I 27 (second century), another letter from Rome written by a grain transporter whose return journey had been delayed; *BGU* II 423 (second century), where the writer thanks Sarapis for saving him from danger at sea.
59 *P. Mich*. VIII 490 (late second century).
60 See Kolb (1997); Thomas (2000); Kolb, this volume and (2000).
61 *P. Oxy*. XXXIV 2719 (third century).
62 *OGIS* 674 = *I. Portes* 80 (AD 90); photograph in Milne (1924) 163 (see Figure 7.3).
63 *O. Claud*. I 58 (second century).
64 Adams (1995) and (1999); Lesquier (1918).
65 *P. Oxy*. XVIII 2128 (AD 165?), with Adams (forthcoming).
66 *P. Oxy*. XXXVI 2760 (*c*. AD 179–80).
67 *BGU* III 665 (first century); *P. Mich*. VIII 492 (second century).
68 *P. Oxy*. XLVI 3313 (second century).
69 Nachtergael (1988).
70 Herodotus II 97; Meyboom (1995).
71 Bowman and Rathbone (1992) 110 n.14.
72 *P. Oxy*. XXXIII 2680 (second or third century); *P. Mich. Zenon* 103, cols 1, 2–8; for roads being declared clear, see *P. Haun*. II 19 (fourth or fifth century).
73 *P. Oxy*. I 118 verso (late third century).
74 *P. Ant*. 43 (late third or early fourth century).
75 Bagnall (1989); McGing (1998).
76 *P. Lille* I 6 (third century).
77 *P. Lond*. VII 2009 (third century BC).
78 *P. Mich*. VI 421 (reign of Claudius), with McGing (1998) 172–3.
79 Bagnall (1977) and (1982).
80 *P. Fuad Univ*. I 26 (AD 59).
81 *P. Lille* II 24 (Ptolemaic period).
82 *P. Oxy*. XIV 1773 (third century).
83 Braunert (1964).
84 Ohler (1986, translation 1989).
85 Macfarlane, cited by Horden and Purcell (2000) 383.
86 *P. Oxy*. XXXIV 2727 (third or fourth century).
87 *P. Brem*. 15 (AD 118).
88 Tacitus, *Annals* II 59.
89 Pliny the Younger, *Epistulae* V 19.
90 *P. Oxy*. X 1271 = *Sel. Pap*. II 304 (AD 264); Strabo 2.3.5.
91 *O. Claud*. I 48–82 (second century).

92 Alston (1998) 189–92. All figures which follow are Alston's.
93 See further, Adams (forthcoming).
94 Alston (1995) 117–42.
95 Hobson (1984); Adams (forthcoming).
96 *P. Ryl.* IV 627–8; *P. Oxy.* XLII 3118; a reference to an itinerary in *P. Mich.*VIII 501.
97 *P. Oxy.* 3052 introduction.

8

AFTERWORD

Travel and empire

Ray Laurence

The traditional consensus on travel was that it was something the élite were involved in, but for the most part others remained in their place of residence as subsistence farmers. Such a view is seriously questioned by the chapters in this volume, which highlight not only the individual acts of travel, but also the complex infrastructure that was necessary to support travel beyond a person's own locale – place of habitation, residence or birth. In part this realisation of the significance of travel comes from the analysis of new evidence, notably papyri from Egypt, that Adams shows point to travel as fundamental in the lives of a set of provincials, whose geography and existence were dominated by the River Nile. On the other hand, new discoveries, such as that of the map of Spain or the *stadiasmus Lyciae*, discussed here by Brodersen and Salway, have been the catalyst for a reappraisal of the nature of ancient geographical knowledge. In all cases, it is found that the nature of this geographical knowledge had a utility for travel with an emphasis on the routes through, rather than description of, areas of territory. It is a mode of expression found in itineraries, and is present in the narrative structures of the Greek novel that is often based on the subject of travel. It may be that what we see within these texts as a journey or narrative is also present in other forms of literature that utilise the activity of travel or journey through life. The use of such metaphors illustrates the significance placed on travel and discovery in the Roman empire as a whole.

The most prominent travellers were those involved in the army. Coulston's study of the frieze on Trajan's Column highlights how the major feature represented is travel across barbarian lands. These places, associated with barbarian peoples and forests, are altered or converted into a landscape organised or regularised by the Roman army. The defeat of enemies appears, at points in the frieze, as a subset of an overall theme of travel from the Danube at the base of the column to the lands of previously unknown barbarians at its summit. This feature of design places an emphasis on the role of the army not just as a pacifier, but also as an explorer of new parts of the globe. A similar structure can be identified in Tacitus' biography of his father-in-law, Agricola. Britons tend to hide in the hills in contrast to the Roman soldiers marching through marshes, across rivers or mountains, led by Agricola, who attempts to woo the Britons with the advantages of towns, large houses, education and the wearing of the toga. Ultimately, Agricola explores Scotland by circumnavigation and campaigns deep into Caledonia resulting in the total defeat and pacification of Britain, according to Tacitus at least. The similarity of representation of military campaigns in these two examples of contemporary evidence points to a close association of the concepts of travel as conquest in texts and visual representations. Both types of evidence do not report accurately the details of campaigns, but instead report the intentions of exploration and pacification of enemies. The scenes that specifically refer to the transportation of soldiers and goods display a variety of methodologies of transport to provide specific examples within the generic structure of representation on the frieze. These are built into the overall structure of war, empire and discovery that would seem to have been the underlying purpose of military campaigns.[1]

The link between geographical knowledge and empire is now widely recognised,[2] but the questions of whether the Romans having maps and the nature of those maps continue to produce debate. Brodersen makes the case that there was little or no need for scale maps, in order to complete or to plan journeys in space and time. The maps that we do possess evidence for are shown by Brodersen and Salway as visual representations of itineraries or lists of places along a route. This discussion takes us down a scale from that of the empire or within a province to the need to travel from place to place, and to know where you were in relationship to other places. Clearly, travel away from the place in which you had been born or lived, beyond the distance of two or three days' journey, to places only heard about as opposed to experienced personally, raises the possibility of becoming lost or never returning. The itinerary or list of places between two major points prevented a person feeling or becoming lost. The traveller using an itinerary would have known which was the next town and which towns they had travelled through to arrive at their destination. Alongside this information was knowledge of distance between places, so that the time taken over a journey may have been estimated. Once in a town on the route, it was a question of asking a local inhabitant which road led to the following town.

The availability of itineraries, and their use, is revealed by Adams's discussion. The surviving evidence points to the usage of itineraries across the empire. On journeys, milestones reminded the traveller of their physical location in relationship to their itinerary with a view to how far they had to travel that day before reaching their next stopping point. In other words, these objects of measurement allowed for the understanding of where you were between points on the major route. They were also utilised for giving directions, for example to turn off to a villa at the such-and-such milestone. Today, when walking and relating our position on the ground to that on a map we may utilise a GPS (global positioning system) in a similar way, but with a greater technological precision. Such accuracy in many ways is unnecessary for most forms of travel leading from A to B as opposed to journeys for recreational hiking or modern warfare.

Intriguingly, alongside milestones, details of itineraries are found also in inscriptions, as Salway highlights. These *tabellaria* are seen to be related to the formation of the itineraries that have come down to us. How necessary it was to use these *tabellaria* comes into question, because if a traveller had an itinerary in a portable form there would be little need for the inscribed versions. However, it may be argued that the *tabellaria*, in particular the *elogium* of Polla, were an earlier, less sophisticated form of measurement of distance that was later superseded by the placing of milestones along the major roads of Italy and the provinces. In fact, anyone wishing to travel from one unknown town to the next need only ask a local inhabitant which was the road to that place. In terms of a parallel, many roads in the towns of Britain are referred to by the name of the place they lead to. There are the following examples in Reading: the Bath Road, the Oxford Road, the Basingstoke Road and the London Road. All these were used to understand the routes to and from a place in the eighteenth and nineteenth centuries. The point to be made here is that itineraries are an efficient way of representing distance and geography, particularly when supplemented by the physical indicators of distance known simply as milestones.

The need to travel would seem to be part of the experience of the Roman empire, which may partially explain its difference from barbarian lands.[3] There is evidence in some form for travel or the need to undertake journeys on the part of individuals from all parts of the empire. Even in the case of Britain, an area incorporated into the empire at a later date than the Mediterranean provinces, there is evidence for the movement of people well beyond the locality of birth or residence – for example, to participate in work on Hadrian's Wall. The relationship of travel to the economy of an individual's family or community remains uncertain. But it is clear from the literature that people travelled over quite large distances for work,[4] and that agricultural practice and building projects depended on migrant labour. The building of monuments in towns often failed due to the absence of skilled labour or may simply not have been finished because the local population was primarily involved in activities that did not allow for time to participate in the construction of monuments.[5] Similarly, travel in order to gain knowledge was a feature of the lives of individuals, whether for a member of the senatorial élite to experience Greek culture at first hand or for a man of medicine to learn about certain cures practised in a particular locality. Indeed, the profession of healing may have been associated with a need for travel from its practitioners' places of origin and residence.[6] Equally, the sick travelled to healing shrines and sanctuaries. All of these individual examples build into an overview of travel within the lives of individuals in the Roman empire. Travel was something that was not an everyday occurrence or an expectation, but most people would have known others who had gone on journeys or had met travellers in their locality.

The difficulty of travel in part comes from a need to rely on facilities and resources at a distance that belong or are organised by other persons. As Adams shows, access to boats or camel trains in parts of Egypt to take a person to a certain place could be limited to an extent where the journey ceased to be possible. The limitations of travel shown here could be seen to provide an insight to the need for the presence of an organised system of transport for imperial officials. Kolb accounts for these factors and provides evidence for the attention to detail in the provision of animals and the use of *mansiones* on the routes used by some officials. The burden of transport and its cost was removed from the imperial exchequer and placed on to local communities for the provision of wagons, transport animals and accommodation along the major routes across the empire. Initially, this system had been set up to make provision for the communication of news from the provinces to the emperor Augustus (whether in Rome or elsewhere), but it quickly became a system that provided for the travel not only of messengers but also of officials such as procurators. The state in effect requisitioned a section of the local economy for the purpose of the provision of transport for state officials. This factor highlights the significance of travel for the functioning of the state and the need to make such journeys more effective by assuring the provision of transport facilities on major routes across the empire.

The complexity of the infrastructure for travel and its importance is highlighted by the administration of Roman law in relationship to roads and transport. The level and scale of the evidence in the *Digest* is at times overwhelming, particularly with reference to the rights of access.[7] When it comes to public roads, managed by either senatorial curators in Italy or governors in the provinces, we find much that indicates the interest of lawyers and the general significance of both the infrastructure and transport itself. The range of matters that the law was concerned with was immense: from the basic definition of a road in terms of width (eight Roman feet), to the formulation that *viae publicae*, like *locae sacrae* or *locae religiosae*, could not be owned, but were the concern of the public.[8] Significantly, a road needed to be wide enough for a vehicle, unlike an *actus* (track) that was suitable for beasts of burden.[9] This causes roads to be very different forms of transport infrastructure, and refers back to the need for transport in a specific form. Indeed, the role of these defined *viae* may be related to the need for the *cursus publicus* to utilise vehicles rather than just pack-animals or riders on horses or mules. The fact that the roads were a public concern causes them to be protected by the law and to prevent persons damaging the road through their private actions – for example, by constructing water pipes or an aqueduct across a road. The permission of the *princeps* or emperor was required before such building work could commence.[10] However, the jurisdiction of the curators of roads and governors of provinces, in the name of the *princeps*, extended beyond the roadway itself to include the maintenance of the *mansiones* and other matters – including the administration of the public land or *ager publicus*.[11] These legal texts point to the administration of the law with reference to the

Figure 8.1 Roman bridge across the River Nar at Narni: a technological innovation to facilitate travel

transport infrastructure, but there is also evidence for the use of the system in terms of the administration of justice. When a person was summoned to court, he needed to travel to the place of the assizes. The lawyers calculated how long it would take a person to undertake this journey and considered that a day was allowed for every twenty miles of travel – a measure that relates distance to the lowest conceivable speed of travel.[12] Other examples include the need for slaves manumitted by a master to travel to the governor of a province to receive their official grant of freedom.[13] There is also within the *Digest* discussion of legal complexities: when another person's slave has been run over due to the negligence of a mule-driver, or when a muleteer has been hired out to another person and through neglect had caused that person's mule to die.[14] This body of jurisprudence points to the complexity and prevalence of transport and travel in the Roman empire. However, this does not mean that journeys were safe or certain: the law makes it clear that death was generally brought on by old age, sickness or bandit attack.[15]

The complexity of the law and the need to organise infrastructure for travel, at a distance from Rome, point to the fundamental importance of travel to the Roman imperial project. The latter is confirmed by the physical survival of Roman roads across Italy and Rome's provinces (Figures 8.1 and 8.2). These structures, in many ways, appear over-engineered to facilitate the movement of people and goods. However, in the recently conquered provinces, such as Britain, the major purpose of such structures may have been to alter the geography of the area and to create a new Roman landscape. This takes us back to Trajan's Column and the activities of soldiers altering the landscape from forest and marshes to those forms that were recognisable as civilised and associated with engineered structures: bridges, towns, forts and so on. The subsequent landscape of a province, as opposed to a land of conquest, was dominated by these long-distance roads that were constructed between places (e.g. legionary fortresses in Britain) to integrate the government and to create a structure of communications within the province. This purpose can be traced back via earlier examples to the building of the via Appia in 312 BC, as forms of long-distance land-based communications. In many ways, it was applied to its most spectacular effect in Spain – a large land mass that did not offer the alternatives of sea transport. The via Augusta in this case linked the Mediterranean to Oceanus (the Atlantic Ocean) over a distance of 1,500 kilometres.[16] The roads as infrastructure altered the nature of geography to create a form of unity and were part of a recognisable landscape that emphasised a cultural unity across the empire in contrast to those regions beyond it.

The government of the empire depended on the mobility of its subjects and its rulers. In the republic, voting at elections and in the assemblies took place in Rome. Therefore, for those living outside the city, participation depended on an ability to travel. Distances could be considerable: for example, the colony at Brindisi was located 360 miles from the capital. There is a tacit assumption in most writing on politics of the republic that individuals from outside Rome

Figure 8.2 A surviving Roman road in Britain: Ackling Dyke. Photograph © Hugh E. Davies

only voted on these occasions when there was a vested interest. However, what is clear is that on matters that were of a concern to them as citizens, they travelled in large numbers and voted. This, in a way, sums up the difference of our concept of representation at elections – voting mattered if it affected your rights as an individual citizen, in which case the journey to Rome was worth the effort. The change of government from a republic to a monarchy, ruled by an emperor, did not end the need for travel to gain representation either in the court or in the senate. What had changed, though, was the mobility of government: the emperor travelled, encountered petitioners and made decisions outside of Rome. This mobility of the court and centre of government had a knock-on effect: information had to find its way to the mobile emperor.

The setting up of the *cursus publicus* by Augustus can be seen as a response to these factors. Government solely from Rome changed to a system of government based mostly in Rome (the senate) with decisions and edicts often made elsewhere by the emperor and his advisers. The flow of members of the Roman élite going to and from the provinces to hold positions in the army from military tribune up to legionary commander points to the experience of travel as a regular occurrence in the life of a member of the élite. Their role as governors, whether in the republic or under the emperors, was one of travel to meet subjects, inspect cities, meet with the army, conduct campaigns against enemies and so on.

The history of Rome is one of ever-increasing complexity as the historian in antiquity grappled with the need to account for the interconnection of events in Rome, on the frontiers and in the provinces. Tacitus' account of the revolt of the legions on the accession of Tiberius reveals the spatial unity of the empire.[17]

Figure 8.3 Representing provinces and peoples: Macedonia, Achaea and the North Africa of the Garamantes on the Peutinger Table (With permission from the Österreichische NationalBibliothek.)

In part, this was dependent on the dispersal of the sons of the imperial family to secure the loyalty of the legions. This dispersed the authority of the emperor across space. When that authority was questioned decisively, the soldiers marched on Rome independently, for example in AD 69.

These major themes of the history and archaeology of the Roman empire can be seen to benefit from an understanding of the nature of travel and geographical knowledge. What we see in the studies in this volume is an understanding of how travel was conducted and how the Roman empire formed a geographical unity, seen, for example, in the Peutinger Table or the Antonine Itineraries (see Figure 8.3). This creates a need to study the archaeology of a Roman province or a city in the provinces, alongside or with reference to other provinces and even the evidence from Italy itself. The twentieth-century tradition developed provincial archaeology of the Roman empire with reference to the modern geography of Europe, the Near East and Africa. This may have created a mode of expression that sees the culture of Roman Britain as distinctive from that of Roman France, in line with early twentieth-century ideologies of empire in the modern world.[18] The provinces of the empire are often studied in isolation without a clear understanding of how each geographical region was connected to Rome. What we have attempted to do in this volume is to find a way to view those connections, whether it is the relationship of a newly conquered province to Rome, the representation of transport and cultural geography on Trajan's Column, or the representation of space and distance. The perspectives in this volume provide us with an insight into areas of study that impact on all the major themes of both Roman history and the archaeology of Rome, Italy and the provinces.

NOTES

1 Compare *Res Gestae* 25–33.
2 For example, Nicolet 1991.
3 Ael. *Arist.* 26.
4 For example, Suet. *Vesp.* 1.
5 For example, Pliny the Younger, *Ep.* 10.37–8.
6 For example, Pliny the Younger, *Ep.* 10.5–6.
7 For example, *Dig.* 7.6.1.3.
8 *Dig.* 8.3.8, 18.1.51.
9 *Dig.* 8.1.13.
10 *Dig.* 39.3.18.
11 *Dig.* 50.4.18.10, 43.23.2.
12 *Dig.* 11.1.1.
13 *Dig.* 12.4.5.2.
14 *Dig.* 9.2.8, 19.2.60.7.
15 *Dig.* 13.6.5.4.

16 For discussion of evidence for roads in the provinces of the empire, see Chevallier (1976) 140–64.
17 *Annals* 1.16–49.
18 See Hingley 2000.

BIBLIOGRAPHY

Abbreviations used for periodicals may be found in *L'Année Philologique* (Paris, 1927–).

Abbott, F.F. and Johnson, A.C. (1926) *Municipal Administration in the Roman Empire* (Princeton).

Adams, C.E.P. (1995) 'Supplying the Roman army: *O Petr.* 245', *ZPE* 109: 119–24.

—— (1999) 'Supplying the Roman army: bureaucracy in Roman Egypt', in A. Goldsworthy and I. Haynes (eds) *The Roman Army as a Community*, JRA Supplementary Series 34 (Ann Arbor), 119–26.

—— (2000) 'Who bore the burden? The organisation of stone transport in Roman Egypt', in D. Mattingly and J. Salmon (eds) *The Productive Past: Economies beyond Agriculture* (London).

—— (forthcoming) *Land Transport in Roman Egypt 30 BC–AD 300: A Study in Administration, Bureaucracy and Economic History* (Oxford).

Adams, J.N. (1977) *The Vulgar Latin of the Letters of Claudius Terentianus (P. Mich. VIII, 467–72)*, Publications of the Faculty of Arts of the University of Manchester, No. 23 (Manchester).

Alföldi, A. (1952) 'The moral frontier on Rhine and Danube', in E. Birley (ed.) *The Congress of Roman Frontier Studies 1949* (Durham), 1–16.

Alfs, J. (1941) 'Der bewegliche Metallpanzer im römischen Heer. Die Geschichte seiner Herkunft und Entwicklung', *Zeitschrift für historische Waffen und Kostumkunde* 3–4: 69–126.

Alston, R. (1994) 'Roman military pay from Caesar to Diocletian', *JRS* 84: 113–23.

—— (1995) *Soldier and Society in Roman Egypt* (London/New York).

—— (1998) 'Trade and the city in Roman Egypt', in H. Parkins and C.J. Smith (eds) *Trade, Traders and the Ancient City* (London).

Ampolo, C. (1995) 'L'*omen victoriae* della Colonna Traiana: il principe e l'uomo caduto dal mulo', *Archeologia Classica* 47: 317–27.

Anderson, J.D. (1992) *Roman Military Supply in North-east England*, BAR British Series 224 (Oxford).

Anderson, J.K. (1985) *Hunting in the Ancient World* (Berkeley).

André, J.-M. and Baslez, M.-F. (1993) *Voyager dans l'Antiquité* (Paris).

Andreae, B. (1979) 'Zum Triumphfries des Trajansbogens von Benevent', *Mitteilungen des Deutschen Archäologischen Instituts, Römische Abteilung* 86: 325–9.

Apianus, P. and Amantius, B. (1534) *Inscriptiones sacrosanctae vetustatis* (Ingolstadt).

Arnaud, P. (1988) 'L'origine, la date de rédaction et la diffusion de l'archetype de la Table de Peutinger', *BSNAF* 302–21.

—— (1989) 'Une deuxième lecture du Bouclier de Doura-Europos', *CRAI*: 373–89.

—— (1992) 'A propos d'un prétendu itinéraire de Caracalla dans l'Itinéraire d'Antonin: les sources tardives de l'itinéraire de Rome à Hierasycaminos', *BSNAF* 374–9.

—— (1993) 'L'*Itinéraire d'Antonin*: un témoin de la littérature itinéraire du Bas-Empire', *Geographia Antiqua* 2: 33–47.

Atlas (1974–9) *Atlas Republica Socialista Romania* (Bucharest).

Austin, N.J.E. and Rankov, N.B. (1995) *Exploratio: Military and Political Intelligence in the Roman World from the Second Punic War to the Battle of Adrianople* (London).

Avi-Yonah, M. (1954) *The Madaba Mosaic Map with Introduction and Commentary* (Jerusalem).

Aymard, J. (1951) *Essai sur les chasses romaines des origines à la fin du siècle des Antonins (Cynogetia)* (Paris).

Baatz, D. (1966) 'Zur Geschützbewaffnung römischer Auxiliartruppen in der frühen und mittleren Kaiserzeit', *Bonner Jahrbücher* 166: 194–207.

—— (1978) 'Recent finds of ancient artillery', *Britannia* 9: 1–17.

—— (1980) 'Ein Katapult der legio IV Macedonica aus Cremona', *Mitteilungen des Deutschen Archäologischen Instituts, Römische Abteilung* 87: 283–99.

—— (1985) 'Hibernacula', *Germania* 63, 147–54 = Baatz 1994, 105–12.

—— (1994) *Bauten und Katapulte des römischen Heeres* (Stuttgart).

Baatz, D. and Bockius, R. (1997) *Vegetius und die römische Flotte* (Mainz).

Baatz, D. and Feugère, M. (1981) 'Eléments d'une catapulte romaine trouvée à Lyon', *Gallia* 39: 201–9.

Bachrach, B.S. (1985) 'Animals and warfare in early medieval Europe', *Settimane di Studio del Centro Italiano sull'alto Medioevo* 31 (= *Armies and Politics in the Early Medieval West* (Aldershot), XVII), 705–51.

Badian, E. (1970) 'Tabellarius', in N.G.L. Hammond and H.H. Scullard (eds) *The Oxford Classical Dictionary*^{2} (Oxford), 1032.

Baehrens, E. (ed.) (1883) *Poetae Latini minores* V (Leipzig).

—— (ed.) (1886) *Fragmenta Poetarum Romanorum* (Leipzig).

Bagnall, R.S. (1969) 'Notes on *P. Hib.* II 198', *BASP* 6: 73–118.

—— (1976) *The Florida Ostraca: Documents from the Roman Army in Upper Egypt* (*GRBS* Monograph 7) (Durham, N.C.).

—— (1977) 'Army and police in Upper Egypt', *JARCE* 14: 67–86.

—— (1982) 'Upper and lower guard posts', *CdÉ* 57: 125–8.

—— (1985) 'The camel, the wagon and the donkey in later Roman Egypt', *BASP* 22: 1–6.

—— (1989) 'Official and private violence in Roman Egypt', *BASP* 26: 201–16.

—— (1993) *Egypt in Late Antiquity* (Princeton).

—— (1995) *Reading Papyri, Writing Ancient History* (London/New York).

Ball, J. (1942) *Egypt in the Classical Geographers* (Cairo).

Barcacila, A. (1966) 'Les piliers du pont Trajan sur la rive gauche du Danube et la scène CI de Colonne Trajan', *Studii si cervetari de istorie veche* 17: 645–63.

Barnes, T.D. (1982) *The New Empire of Diocletian and Constantine* (Harvard).

Barrett, A.A. (1979) 'The Career of Cogidubnus', *Britannia* 10: 203–26.

Bartoli, P.S. (1672) *Colonna Traiana, eretto dal Senato e Popolo Romano all'imperatore Traiano Augusto nel suo foro in Roma* (Rome).

Basch, L. (1987) *Le musée imaginaire de la marine antique* (Athens).

Baumer, L.E., Hölscher, T. and Winkler, L. (1991) 'Narrative systematik und politisches Konzept in den Reliefs der Traianssäule. Drei Fallstudien', *Jahrbuch des deutschen Archäologischen Instituts* 106, 261–95.

Becatti, G. (1961) *Scavi di Ostia IV. Mosaici e Pavimenti Marmorei* (Rome).

Bejaoui, F. (1999) 'Découverte dans l'antique Haidra: la Méditerranée sur une mosaique', *Archéologie* 357: 16–23.

—— (1999/2000) 'Iles et villes de la Méditerranée sur une mosaique d'Ammaedara', *Comptes rendus de l'Académie des Inscriptions et Belles-Lettres 1997*, 827–60.

Bekker-Nielsen, T. (1988) 'Terra Incognita: the subjective geography of the Roman empire', in A. Dansgaard-Madsen, E. Christiansen and E. Hallger (eds) *Studies in Ancient History and Numismatics Presented to Rudi Thomsen* (Århus), 148–61.

—— (1989) *The Geography of Power. Studies in the Urbanization of Roman North-west Europe*, BAR International Series 477 (Oxford).

Belloni, G.G. (1973) *Le Monete di Traiano* (Milan).

Bender, H. (1978) *Römischer Reiseverkehr. Cursus publicus und Privatreisen* (Aalen).

Berchem, D. van (1937) 'L'annone militaire dans l'empire romain au IIIe siècle', *MSNAF* 80: 117–202.

—— (1973) 'Les itinéraires de Caracalla et l'Itinéraire d'Antonin', *CRAI*: 123–6.

—— (1974) 'Les itinéraires de Caracalla et l'Itinéraire d'Antonin', in *Actes du IXe Congrès international d'études sur les frontières romaines, Mamaia, 6–13 Sept. 1972* [*Limes Congress IX*], Editions de l'Académie (Cologne), 301–7.

Bertinetti, M., Lachenal, L. and De Palma, B. (eds) (1985) *Museo Nazionale Romano, Le Sculture* I.8.i (Rome).

Bishop, M.C. (1988) 'Cavalry equipment of the Roman army in the first century AD', in Coulston 1988a, 67–195.

Bishop, M.C. and Coulston, J.C.N. (1993) *Roman Military Equipment from the Punic Wars to the Fall of Rome* (London).

Black, E.W. (1995) *Cursus Publicus. The Infrastructure of Government in Roman Britain*, BAR British Series 241 (Oxford).

Blackman, D.J. (1982a) 'Ancient harbours in the Mediterranean. Part 1', *International Journal of Nautical Archaeology and Underwater Exploration* 11.2: 79–104.

—— (1982b) 'Ancient harbours in the Mediterranean. Part 2', *International Journal of Nautical Archaeology and Underwater Exploration* 11.3: 185–211.

Blyth, P.H. (1992) 'Apollodorus of Damascus and the *Poliorcetica*', *GRBS* 33: 127–58.

Boardman, J. (1978) *Greek Sculpture. The Archaic Period* (London).

—— (1985) *Greek Sculpture. The Classical Period* (London).

Bogaers, J.E. (1979) 'King Cogidubnus: Another Reading of *RIB* 91', *Britannia* 10: 243–54.

Bökönyi, S. (1988) 'Animal breeding on the Danube', in C.R. Whittaker (ed.) *Pastoral Economies in Classical Antiquity* (Cambridge), 171–6.

Bosio, L. (1983) *La Tabula Peutingeriana.Una descrizione pittorica del mondo antico* (I monumenti del arte classica 2) (Rimini).

Bowman, A.K. (1971) *The Town Councils of Roman Egypt* (Toronto).

Bowman, A.K. and Rathbone, D. (1992) 'Cities and administration in Roman Egypt', *Journal of Roman Studies* 82:107–27.

Bowman, A.K. and Woolf, G. (eds) (1994) *Literacy and Power in the Ancient World* (Cambridge).

Boymel Kampen, N. (1995) 'Looking at gender: the Column of Trajan and Roman historical relief', in D.C. Stanton and A.J. Stewart (eds) *Feminisms in the Academy* (Ann Arbor), 46–73.

Bracco, V. (1985) 'Il *tabellarius* di Polla', *Epigraphica* 47: 93–7.

Braund, D. (1996) 'River frontiers in the environmental psychology of the Roman world', in D.L. Kennedy (ed.) *The Roman Army in the East*, Journal of Roman Archaeology Supplementary Series 18 (Ann Arbor), 43–7.

Braunert, H. (1964) *Die Binnenwanderung – Studien zur Sozialgeschichte Ägyptens in der Ptolemäer- und Kaiserzeit* (Bonn).

Brett, G. (1947) *The Great Palace of the Byzantine Emperors* (Oxford).

Brilliant, R. (1967) *The Arch of Septimius Severus in the Roman Forum*, Memoirs of the American Academy at Rome 29 (Rome).

—— (1984) *Visual Narratives, Storytelling in Etruscan and Roman Art* (Ithaca).

Brodersen, K. (1995) *Terra Cognita, Studien zur römischen Raumerfassung*, Spudasmata 59 (Zurich and New York).

—— (1996) 'Principia geographiae: Antike Texte im fruehen Erdkunde-Unterricht', *Anregung* 42: 29–43.

—— (1996/97) 'Miliarium aureum und Umbilicus Romae: Zwei Mittelpunkte des römischen Reiches?', *Wuerzburger Jahrbuecher* N.F. 21: 273–83.

—— (1997) 'Artemidoros 3', *Der Neue Pauly II* (Stuttgart): 50.

—— (1999) 'Mapping the ancient world', *Ad familiares: Journal of the Friends of Classics* 17: 2–4.

—— (2000) *Mastering the World: Ancient Geography* (London).

—— (2001) 'Neue Entdeckungen zu antiken Karten', *Gymnasium* 108: 137–48.

—— (forthcoming) 'Die Tabula Peutingeriana: Gehalt und Gestalt einer "alten Karte" und ihrer antiken Vorlage', in D. Unverhau (ed.) *Geschichtsdeutung auf alten Karten*, Wolfbenbütteler Forschungen (Wiesbaden).

Brunt, P.A. and Moore, J.M. (eds) (1967) *Res Gestae Divi Augusti: The Achievements of the Divine Augustus* (Oxford).

Buck, R.J. (1983) *Agriculture and Agricultural Practice in Roman Law*, Historia Einzelschriften 45 (Wiesbaden).

Bühl, G. (1995) *Constantinopolis und Roma: Stadtpersonifikationen der Spätantike* (Crescens 3) (Zürich).

Bulliet, R.W. (1975) *The Camel and the Wheel* (Cambridge, MA).

Bülow, G. von (1980) 'Thrakerdarstellungen in der römischen Kunst', *Klio* 62: 139–55.

Burford, A. (1960) 'Heavy transport in classical antiquity', *Economic History Review* 13: 1–18.

Burnham, B.C. (1979) 'Pre-Roman and Romano-British urbanism? Problems and possibilities', in B.C. Burnham and H.B. Johnson (eds) *Invasion and Response: The Case of Roman Britain*, BAR British Series 73 (Oxford), 255–72.

Burnham, B.C. and Wacher, J. (1990) *The 'Small Towns' of Roman Britain* (London).

Calabi Limentani, I. (1968) *Epigrafia latina*[2] (Biblioteca storica universitaria, serie 1, trattati, 3) (Milan and Varese).

Campbell, D.B. (1986) 'Auxiliary artillery revisited', *Bonner Jahrbücher* 186: 117–32.
Camus, P. (1974) *Le Pas des légions* (Paris).
Capelli, R. and Pesando, F. (1991) 'Gli itinerari romani. Repertorio bibliografico', in R. Capelli (ed.) *Viae Publicae Romanae: X Mostra europea del Turismo, artigianato e delle tradizioni culturali, Roma, Castel Sant'Angelo, 11–25 aprile 1991* (Rome), 41–4.
Cary, M. (1936) 'Direction-posts on Roman roads?', *CR* 50: 166–7.
Casey, P.J. (1994) *Carausius and Allectus: The British Usurpers* (London).
Casson, L. (1965) 'Harbour and river boats of ancient Rome', *Journal of Roman Studies* 55: 31–9.
—— (1971) *Ships and Seamanship in the Ancient World* (Princeton).
—— (1974) *Travel in the Ancient World* (London).
—— (1994) *Ships and Seafaring in Ancient Times* (London).
Chandler, D. (1976) *The Art of Warfare in the Age of Marlborough* (London).
Chevallier, R. (1972) *Les voies romaines* (Collection U, série Histoire ancienne) (Paris).
—— (1976) *Roman Roads* (London).
Chevedden, P.E. (1995) 'Artillery in Late Antiquity: prelude to the Middle Ages', in I.A. Corfis and M. Wolfe (eds) *The Medieval City under Siege* (Woolbridge), 131–73.
Cichorius, C. (1896–1900) *Die Reliefs der Traianssäule*, II–III (Berlin).
Cisek, E. (1983) *L'epoque de Trajan. Circonstances politiques et problèmes idéologiques* (Bucharest).
Clairmont, C.W. (1975) *Excavations at Salona (1969–1972)* (Park Ridge).
Claridge, A. (1993) 'Hadrian's Column of Trajan', *Journal of Roman Archaeology* 6: 5–22.
Clauss, M. (1973) *Untersuchungen zu den principales des römischen Heeres von Augustus bis Diokletian. Cornicularii, speculatores, frumentarii* (Bochum).
—— (1980) *Der Magister officiorum in der Spätantike* (Munich).
Colonna (1988) *La Colonna Traiana e gli artisti francesi da Luigi XIV a Napoleone I* (Rome).
Connolly, P. (1981) *Greece and Rome at War* (London).
—— (1987) 'The Roman saddle', in M. Dawson (ed.) *Roman Military Equipment: The Accoutrements of War. Proceedings of the Third Roman Military Equipment Research Seminar*, BAR International Series 336 (Oxford), 7–27.
Conole, P. and Milns, R.D. (1983) 'Neronian frontier policy in the Balkans: the career of Ti. Plautius Silvanus', *Historia* 32: 183–200.
Cosme, P. (1998) *L'état romain entre éclatement et continuité* (Histoire, cultures et sociétés) (Paris).
Couissin, P. (1926) *Les armes romaines* (Paris).
Coulston, J.C.N. (1985) 'Roman archery equipment', in M.C. Bishop (ed.) *The Production and Distribution of Roman Military Equipment. Proceedings of the Second Roman Military Equipment Research Seminar* (Oxford), 220–366.
—— (1986) 'Roman, Parthian and Sassanid tactical developments', in P. Freeman and D. Kennedy (eds) *The Defence of the Roman and Byzantine East*, BAR International Series 297 (Oxford), 59–75.
—— (ed.) (1988a) *Military Equipment and the Identity of Roman Soldiers. Proceedings of the Fourth Roman Military Conference*, BAR International Series 394 (Oxford).
—— (1988b) 'Three legionaries at Croy Hill', in Coulston 1988a, 1–29.
—— (1989) 'The value of Trajan's Column as a source for military equipment', in C. van Driel-Murray (ed.) *Roman Military Equipment: The Sources of Evidence.*

Proceedings of the Fifth Roman Military Equipment Conference, BAR International Series 476 (Oxford), 31–44.

Coulston, J.C.N. (1990a) 'Three new books on Trajan's Column', *Journal of Roman Archaeology* 3: 290–309.

—— (1990b) 'The architecture and construction scenes on Trajan's Column', in M. Henig (ed.) *Architecture and Architectural Sculpture in the Roman Empire* (Oxford), 39–50.

—— (forthcoming (a)) 'Emperor and army on Trajan's Column: a study in composition and propaganda', in A.M. Liberati (ed.) *Traiano. Optimus Princeps* (Rome).

—— (forthcoming (b)) 'Tacitus, *Historiae* I.79 and the impact of Sarmatian warfare on the Roman empire', in C. von Carnap-Bornheim (ed.) *Kontakt – Kooperation – Konflikt: Germanen und Sarmaten zwischen dem 1. und 4. Jahrh. n. Chr.* (Marburg).

Crefeld, H. van (1977) *Supplying War. Logistics from Wallenstein to Patten* (Cambridge).

Creighton, J. (2000) *Coins and Power in Late Iron Age Britain* (Cambridge).

Crogiez, S. (1990) 'Les stations du *cursus publicus* en Calabrie: un état de la recherche', *MEFRA* 102: 389–431.

Cumont, F. (1926) *Fouilles de Doura-Europos*, 2 vols (Paris).

Cuntz, O. (1893) 'Beiträge zur Textkritik des Itinerarium Antonini', *Wiener Studien* 15: 260–98.

—— (ed.) (1929) *Itineraria Romana* vol. I: *Itineraria Antonini Augusti et Burdigalense* (Leipzig) (repr. with updated bibliography by G. Wirth (Stuttgart, 1990)).

Dabrowa, E. (1991) '*Dromedarii* in the Roman army: a note', in Maxfield and Dobson, 364–6.

d'Ambra, E. (1998) *Art and Identity in the Roman World* (London).

Darwell-Smith, R.H. (1996) *Emperors and Architecture: A Study of Flavian Rome* (Brussels).

Davies, G.A.T. (1917) 'Trajan's First Dacian War', *Journal of Roman Studies* 7: 74–97.

—— (1920) 'Topography and the Trajan's Column', *Journal of Roman Studies* 10: 1–28.

Davies, P.J.E. (1997) 'The politics of perpetuation: Trajan's Column and the art of commemoration', *American Journal of Archaeology* 101: 41–65.

Davies, R.W. (1967) '*Ratio* and *opinio* in Roman military documents', *Historia* 16: 115–18.

—— (1969a) 'Joining the Roman army', *Bonner Jahrbücher* 169: 208–32 = Davies (1989) 2–30.

—— (1969b) 'The supply of animals to the Roman army and the remount system', *Latomus* 28: 429–59 = Davies (1989) 152–73.

—— (1971) 'The Roman military diet', *Britannia* 2: 122–42 = Davies (1989) 187–206.

—— (1989) *Service in the Roman Army* (Edinburgh).

Davison, D.P. (1989) *The Barracks of the Roman Army from the 1st to 3rd Centuries AD*, BAR International Series 472 (Oxford).

De Rosalia, A. (ed.) (1978) *Iscrizioni latine arcaiche*[2] (Hermes–Collana di testi antichi 8) (Palermo).

Degrassi, A. (1937) *Inscriptiones Italiae* 13, fasc. 2, Istituto Poligrafico dello Stato (Rome).

—— (1957) *Inscriptiones Latinae Liberae Rei Publicae* I (Biblioteca di studi superiori 23, sezione Storia antica ed epigrafia) (Florence).

—— (1965) *Inscriptiones Latinae Liberae Rei Publicae: Imagines* (Berlin).

—— (1986) *Corpus Inscriptionum Latinarum* 1, 2nd edn, part 2, fasc. iv (Berlin).

Degrassi, A. and Krummrey, J. (eds) (1986) *Corpus Inscriptionum Latinarum* vol. I^2, *Inscriptiones Latinae Antiquissimae*, part 2, fasc. lv (Berlin).

Delmaire, R. (1989) *Largesses sacrées et res privata. L'aerarium impérial et son administration du IV*[e] *au VI*[e] *siècle* (Rome).

Deman, A. (1976) '*Tabelarios*. L'inscription routière de C. Popillius Laenas et une ordonnance de Jean Ier ou Jean II le Bon', in G. Cambier (ed.) *Hommages à André Boutemy* (Collection Latomus, vol. 145) (Brussels), 32–7.

Demandt, A. (1989) *Die Spaetantike: Römische Geschichte von Diocletian bis Justinian 284–565 n.Chr.* Handbuch der Altertumswissenschaft III 6. (Munich).

Demougeot, E. (1984) 'L'image officielle du barbare dans l'empire romain d'Auguste à Théodose', *Ktema* 9: 123–43.

Desjardins, E.E.A. (1876–93) *Géographie historique et administrative de Gaule*, 4 vols (Paris).

—— (1878) *Les Tabellarii* (Bibliothèque de l'Ecole des hautes études, Section des sciences philologique et historiques, 35) (Paris).

Dessau, H. (1929) 'Mommsen und das Monumentum Ancyranum', *Klio* 22: 261–83.

Detlefsen, S.D.F. (1909) *Die Anordnung der geographischen Bücher des Plinius und ihre Quellen* (Quellen und Forschungen zur alten Geschichte und Geographie 18) (Berlin) (repr. Rome, 1972).

Dilke, O.A.W. (1985) *Greek and Roman Maps* (London).

—— (1987) 'Itineraries and geographical maps in the early and late Roman empires', in J.B. Harley and D. Woodward (eds) *The History of Cartography*, I, *Cartography in Prehistoric, Ancient, and Medieval Europe and the Mediterranean* (Chicago), 234–57.

—— (1988) 'Table ronde on Graeco-Roman cartography (Paris 1987)' *Journal of Roman Archaeology* 1: 89–94.

Dillemann, L. (1997) *La Cosmographie du Ravennate* (Collection Latomus 235) (Brussels).

Doblhofer, E. (ed.) (1972–7) *Rutilius Claudius Namatianus, De reditu suo siue Iter Gallicum* (Wissenschaftliche Kommentare zu griechischen und lateinischen Schriftstellern), 2 vols (Heidelberg).

Donner, H. and Cueppers, H. (1977) *Die Mosaikkarte von Madeba I*: Tafelband. Abhandlungen des deutschen Palaestinavereins (Wiesbaden).

Douglass, L. (1996) 'A New Look at the *Itinerarium Burdigalense*', *JECS* 4: 313–33.

Driel-Murray, C. van (1991a) 'A Roman tent: Vindolanda Tent I', in Maxfield and Dobson, 367–72.

—— (1991b) 'New light on old tents', *Journal of Roman Military Equipment Studies* 1: 109–37.

Drower, M.S. (1985) *Flinders Petrie: A Life in Archaeology* (London).

Eck, W. (1979) *Die staatliche Organisation Italiens in der hohen Kaiserzeit* (Munich).

—— (1992) 'Supermentarii et muliones im privaten eines römischen statthalters' *ZPE* 90: 207–10.

—— (1995) 'Zur Durchsetzung von Anordnungen und Entscheidungen in der hohen Kaiserzeit: Die Administrative Informationsstruktur', in R. Frei-Stolba and M.A. Speidel (eds) *Die Verwaltung des römischen Reiches in der hohen Kaiserzeit. Ausgewählte und erweiterte Beiträge*. 1. Band, Arbeiten zur römischen Epigraphik und Altertumskunde, Band 1.

Eck, W. (1997–8) 'Tacitus, Ann. 4.27.1 und der cursus publicus auf der Adria', in R. Frei-Stolba and M. Speidel (eds) *Die Verwaltung des römischen Reiches in der hohen Kaiserzeit*. Ausgewälte und erweiterte Beiträge. 2. Band, Arbeiten zur römischen Epigraphik und Altertumskunde, Band 3, 339–46.

Elter, A. (1908) *Itinerarstudien* vol. I (Bonn).

Engels, D.W. (1978) *Alexander the Great and the Logistics of the Macedonian Army* (Berkeley).

—— (1999) *Augusteische Oikumenegeographie und Universalhistorie im Werk Strabons von Amaseia* (Geographica Historica 12) (Stuttgart).

Erdkamp, P. (1998) *Hunger and the Sword. Warfare and Food Supply in the Roman Republican Wars (264–30 BC)* (Amsterdam).

Espérandieu, E. (1907–81) *Recueil général des bas-reliefs, statues et bustes de la Gaule romaine* (Paris).

—— (1931) *Recueil général des bas-reliefs, statues et bustes de la Germanie romaine* (Paris).

Farinella, V. (1981) 'La Colonna Traiana: un esempio di lettura verticale', *Prospettiva* 26: 2–9.

Fink, R.O. (1971) *Roman Military Records on Papyrus*, American Philological Association Monograph 26 (Cleveland).

Finkel, C. (1988) *The Administration of Warfare: The Ottoman Military Campaigns in Hungary, 1593–1606* (Vienna).

Finley, M. (1973) *The Ancient Economy* (London).

Flaubert, G. (1983) *Flaubert in Egypt* (tr. F. Steegmuller) (Chicago).

Florescu, F.B. (1965) *Das Siegesdenkmal von Adamklissi Tropaeum Traiani* (Bonn).

Flower, H.I. (1996) *Ancestor Masks and Aristocratic Power in Roman Culture* (Oxford).

Foertmeyer, V.A. (1989) *Tourism in Graeco-Roman Egypt*, unpublished Ph.D. thesis, University of Princeton.

Fradier, G. (1989) *Mosaïques Romaines de Tunisie* (Tunis).

Franken, N. (1999) 'Elefantreiter. Zum Typus der Barbarenstatuette aus Großsachsenheim', *Jahrbuch des Deutschen Archäologischen Instituts* 114, 125–56.

Freccia, M. (1554) *De subfeudis baronum, et inuestituris feudorum* (Naples) (rev. edn Venice, 1579).

French, D.H. (1996) 'The site of Baratha and routes on the Konya Plain', *EA* 27: 93–101.

Frenz, H.G. (1992) *Corpus Signorum Imperii Romani, II.7, Germania Superior. Bauplastik und Porträts aus Mainz und Umgebung* (Mainz).

Frere, S.S. (1974) *Britannia* (London).

Freshfield, E.H. (1921–2) 'Notes on a *vellum album* containing some original sketches of public buildings and monuments, drawn by a German artist who visited Constantinople in 1574', *Archaeologia* 72: 87–104.

Friedländer, L. (1908–13) *Roman Life and Manners under the Early Roman Empire* 4 vols (London).

Fuentes, N. (1991) 'The mule of a soldier', *Journal of Roman Military Equipment Studies* 2: 65–99.

Futrell, A. (1997) *Blood in the Arena. The Spectacle of Roman Power* (Austin).

Gall, H. von (1990) *Das Reiterkampfbild in der iranischen und iranisch beeinflussten Kunst parthischer und sasanidischer Zeit* (Berlin).

Gallazzi, C. and Kramer, B. (1998/9) 'Artemidor im Zeichensaal. Eine Papyrusrolle mit Text, Landkarte und Skizzenbüchern aus späthellenistischer Zeit', *APF* 44: 190–208.

Gardiner, R. (ed.) (1995) *The Age of the Galley. Mediterranean Oared Vessels since Pre-classical Times* (London).

Garland, K. (1994) *Mr Beck's Underground Map* (Harrow Weald).

Gauer, W. (1977) *Untersuchungen zur Traianssäule*, Monumenta Artis Romae 13 (Berlin).

Geisberg, M. (1974) *The German Single-leaf Woodcut, 1500–1550* (rev. and ed. W.L. Strauss (New York)).

Giardina, A. (1977) *Aspetti della burocrazia nel basso impero* (Rome).

Gilliam, J.F. (1965) 'Dura rosters and the *constitutio Antoniniana*', *Historia* 14: 74–92.

Gilliver, C.M. (1999) *The Roman Art of War* (London).

Giuliano, A. (ed.) (1981) *Museo Nazionale Romano, Le sculture*, I.2 (Rome).

Givigliano, G.P. (1994) 'Percorsi e strade', in S. Settis (ed.) *Storia della Calabria antica* II. *Età italica e romana* (Rome and Reggio Calabria), 243–362.

Glare, P.G.W. (ed.) (1982) *Oxford Latin Dictionary*, fasc. VIII (Oxford).

Glodariu, I. (1974) 'Itinéraires possibles de la cavalerie maure dans les guerres Daces', in H. Daicoviciu (ed.) *In Memoriam C. Daicoviciu* (Cluj), 151–64.

—— (1976) *Dacian Trade with the Hellenistic and Roman World*, BAR Supplementary Series 8 (Oxford).

Glorie, F. (1965) *Itineraria et alia Geographica* (Corpus Christianorum, series Latina 175) (Turnhout).

Goldsworthy, A.K. (1996) *The Roman Army at War, 100 BC–AD 200* (Oxford).

Greene, K. (1986) *The Archaeology of the Roman Economy* (London).

Grenier, A. (1931–60) *Manuel d'archéologie gallo-romaine*, 4 vols (Paris).

Gross, W.H. (1975) 'Tabellarius', in K. Ziegler and W. Sontheimer (eds) *Der Kleine Pauly* vol. V (Munich), cols 475–6.

Habicht, C. (1958) *RE* VIII A2 (Stuttgart), 1928–40.

—— (1985) *Pausanias' Guide to Ancient Greece* (Sather Classical Lectures, vol. 50) (Berkeley and Los Angeles).

Haldon, J. (1999) *Warfare and Society in the Byzantine World, 565–1204* (London).

Hannestad, N. (1979) 'Rome – ideology and art. Some distinctive features', in M. Sordi (ed.) *Conoscenze etniche e rapporti di convivenza nell'antichità* (Milan), 361–90.

—— (1986) *Roman Art and Imperial Policy* (Århus).

Harley, J.B. and Woodward, D. (eds) (1987) *The History of Cartography I: Cartography in Prehistoric, Ancient, and Medieval Europe and the Mediterranean* (Chicago and London).

Harvey, P.D.A. (1987) 'Local and regional cartography in medieval Europe', in Harley and Woodward 1987, 464–501.

Hassel, F.J. (1966) *Der Trajansbogen in Benevent. Ein Bauwerk des römischen Senates* (Mainz).

Haverfield, F. (1912) *The Romanization of Britain* (Oxford).

—— (1924) *The Roman Occupation of Britain* (Oxford).

Herz, P. (1988) *Studien zur römischen Wirtschaftsgesetzgebung* (Stuttgart).

Heurgon, J. (1951) 'The Amiens Patera', *Journal of Roman Studies* 41: 22–4.

—— (1952a) 'La date des goblets de Vicarello', *REA* 54: 39–50.

Heurgon, J. (1952b) 'La fixation des noms de lieux en latin d'après les itinéraires routiers', *RPh* 3e série 26: 169–78.

Hingley, R. (2000) *Roman Officers and English Gentlemen. The Imperial Origins of Roman Archaeology* (London).

Hinrichs, F.T. (1987) 'Review of Dilke 1985', *Historische Zeitschrift* 244: 665–9.

Hirschfeld, O. (1907) 'Die römischen Meilensteine', *SBAW* 165–201.

—— (1913) 'Die agentes in rebus', *Kleine Schriften* 576–612.

—— (1963) *Die kaiserlichen Verwaltungsbeamten bis auf Diokletian* (Berlin).

Hobson, D. (1984) 'Agricultural land and economic life in Socnopaiou Nesos', *BASP* 21: 89–109.

Höckmann, O. (1995) 'Antike Schiffsfunde aus der Donau', in *In Poseidons Reich. Archäologie unter Wasser*, Zaberns Bildbände zur Archäologie 23 (Mainz), 82–90

Holmberg, E.J. (1933) *Zur Geschichte des Cursus Publicus* (Diss., Uppsala).

Hölscher, T. (1980) 'Die Geschichtsauffassung in der römischen Repräsentationskunst', *Jahrbuch des Deutschen Archäologischen Instituts* 95, 265–321.

Hope, V. (1998) 'Negotiating identity and status: the gladiators of Roman Nimes', in R. Laurence and J. Berry (eds) *Cultural Identity in the Roman Empire* (London), 179–95.

Horden, P. and Purcell, N. (2000) *The Corrupting Sea: A Study of Mediterranean History* (Oxford).

Horn, H.G. and Rüger, C.B. (eds) (1979) *Die Numider, Reiter und Könige nördlichen der Sahara* (Cologne).

Hudemann, E.E. (1878) *Geschichte des römischen Postwesens während der Kaiserzeit*, 2nd edn (Berlin).

Hughes, J.D. (1994) *Pan's Travail. Environmental Problems of the Ancient Greeks and Romans* (Baltimore).

Humbert, G. (1887) 'Cursus publicus', in Ch. Daremberg and E. Saglio (eds) *Dictionaire des Antiquités grecques et romaines* I.2 (Paris), 1648.

Hunt, E.D. (1982) *Holy Land Pilgrimage in the Later Roman Empire* AD *312–460* (Oxford).

Hyland, A. (1990) *Equus. The Horse in the Roman World* (London).

Isik, F. (1999) 'Patara, eine Lykische Metropole erwacht aus ihrem 'Dornroeschenschlaf', *Antike Welt* 30: 477–93.

Isserlin, R. (1998) 'A spirit of improvement? Marble and the culture of Roman Britain', in R. Laurence and J. Berry (eds) *Cultural Identity in the Roman Empire* (London), 125–55.

James, S. (1983) 'Archaeological evidence for Roman incendiary projectiles', *Saalburg Jahrbuch* 39: 142–3.

Janni, P. (1984) *La mappa e il periplo: Cartografia antica e spazio odologico* (Università di Macerata, Pubblicazioni della Facoltà di Lettere e Filosofia, 19) (Rome).

Jobst, W. and Weber, E. (1989) 'Ein heiligtum der pferdegötten epona in Carnuntum', *A. Arch. Hung.* 41: 349–58.

Johnson, A. (1983) *Roman Forts of the 1st and 2nd Centuries* AD *in Britain and the German Provinces* (London).

Johnson, A.C. (1936) *Roman Egypt from Augustus to Diocletian* = Frank, T. (ed.) *An Economic Survey of the Roman World,* vol. II (Baltimore).

Johnston, A.E.M. (1967) 'The earliest preserved Greek map: a new Ionian coin type', *JHS* 87: 86–94.

Jones, A.H.M. (1964) *The Later Roman Empire 284–602* (Oxford).
Jones, G.D.B. (1984) "Becoming different without knowing it': the role and development of *vici*', in T. Blagg and A. King (eds) *Military and Civilian in Roman Britain*, BAR British Series 136 (Oxford), 75–91.
Jones, G.D.B. and Mattingly, D. (1990) *An Atlas of Roman Britain* (Oxford).
Junkelmann, M. (1986) *Die Legionen des Augustus. Der römische Soldat im archäologischen Experiment* (Mainz).
—— (1990) *Die Reiter Roms* I, *Reise, Jagd, Triumph und Circusrennen* (Mainz).
—— (1992) *Die Reiter Roms* III, *Zubehör, Reitweise, Bewaffnung* (Mainz).
—— (1997) *Panis Militaris. Die Ernährung des römischen Soldaten oder der Grundstoff der Macht* (Mainz).
Kennedy, D. (1984) 'The composition of a military work-party in Roman Egypt (*ILS* 2483: Coptos)', *Journal of Egyptian Archaeology* 70: 156–60.
Keppie, L.J.F. (2000) *Understanding Roman Inscriptions*, 2nd edition (London and New York).
Keppie, L.J.F. and Arnold, B.J. (1984) *Corpus Signorum Imperii Romani, Great Britain*, I.4, *Scotland* (Oxford).
King, A. (1984) 'Animal bones and the dietary identity of military and civilian groups in Roman Britain, Gaul and Germany', in T.F.C. Blagg (ed.) *Military and Civilian in Roman Britain: Cultural Relations in a Frontier Province*, BAR British Series 136 (Oxford), 187–217.
—— (1999) 'Animals and the Roman army: the evidence of animal bones', in A. Goldsworthy and I. Haynes (eds) *The Roman Army as a Community* (Portsmouth, RI), 139–49.
Kirk, W. (1979) 'The making and impact of the British Imperial North-West Frontier in India', in B.C. Burnham and H.B. Johnson (eds) *Invasion and Response. The Case of Roman Britain*, BAR British Series 73 (Oxford), 39–55.
Kissel, T.K. (1995) *Untersuchungen zur Logistik des römischen Heeres in den Provinzen des griechischen Ostens (27 v. Chr. – 235 n. Chr.)* (St Katherinen).
Kleiner, F.S. (1991) 'The trophy on the bridge and the Roman triumph over nature', *L'Antiquité Classique* 60: 182–92.
Klotz, A. (1931) 'Die geographischen commentarii des Agrippa und ihre Uberreste', *Klio* 24: 38–58, 386–466.
Knauer, E.R. (1998) *The Camel's Load in Life and Death* (Zürich).
Koeppel, G.M. (1980) 'A military itinerarium on the Column of Trajan', *Mitteilungen des deutchen Archäologischen Instituts, Römische Abteilungen*, 301–6.
—— (1982) 'The grand pictorial tradition of Roman historical representation during the early empire', *Aufstieg und Niedergang der römischen Welt* II, *Kunste* 12.1, 507–35.
—— (1985) 'Die historischen Reliefs der römischen Kaiserzeit, III. Stadtrömische Denkmäler unbekannter Bauzugehörigkeit aus trajanischer Zeit', *Bonner Jahrbücher* 185: 143–213.
Kolb, A. (1997) 'Der *cursus publicus* in Aegypten', *Akten des 21. Internationalen Papyrologenkongresses, Berlin 1995* (*APF* 3), 533–40.
—— (2000) *Transport und Nachrichtentransfer im Römischen Reich* (Berlin).
Kortepeter, C.M. (1973) *Ottoman Imperialism during the Reformation: Europe and the Caucasus* (London).
Kramer, B. (1986) 'Zwei Leipziger papyri', *APF* 32: 33–9.

Kramer, B. and Kramer, J. (2000) 'Iberia, Hispania und das neue Artemidor-Fragment', in A. Haltenhoff and F.-H. Mutschler (eds) *Hortus litterarum antiquarum: Festschrift für Hans Armin Gärtner* (Heidelberg), 309–22.

Kromayer, J. and Veith, G. (1923) *Heerwesen und Kriegführung der Griechen und Römer* (Munich).

Krüger, M. (1970) *Corpus Signorum Imperii Romani, Osterreich*, I.3, *Die Skulpturen des Stadtgebietes von Carnuntum* 1, *Die Figurlichen Reliefs* (Vienna).

Kubitschek, J.W. (1891) 'Zur Kritik des Itinerarium Antonini', *WS* 13: 177–209.

—— (1902) 'Eine römische Straßenkarte', *JOAI* 5: 81.

—— (1914) 'Itinerarien', *Real-Encyclopädie der Altertumswissenschaft* vol. IX, cols 2308–63.

—— (1919a) *Itinerar-Studien* (Akademie der Wissenschaften in Wien, Philosophisch-historische Klasse, Denkschriften, 61. Band, 3. Abhandlung) (Vienna).

—— (1919b) 'Karten', *Real-Encyclopädie der Altertumswissenschaft* vol. X, cols 2022–149.

Kuhnen, H.P. (1991) 'Der sarazenensattel: zu den Voraussetzungen der Sarazeneneinfälle am Limes Arabiae', in Maxfield and Dobson, 326–34.

Kunkel, W. and Wittmann, R. (1995) *Staatsordnung und Staatspraxis der Römischen Republik* (Munich).

Lancaster, L. (1999) 'Building Trajan's Column', *American Journal of Archaeology* 103: 419–39.

Lasserre, F. (ed.) (1967) *Strabon, Géographie* III (Paris).

Laubscher, H.P. (1975) *Der Reliefschmuck des Galeriusbogens in Thessaloniki* (Berlin).

Laurence, R. (1999) *The Roads of Roman Italy: Mobility and Cultural Change* (London).

Leander Touati, A.M. (1987) *The Great Trajanic Frieze. The Study of a Monument and of the Mechanisms of Message Transmission in Roman Art* (Stockholm).

Lehmann-Hartleben, K. (1926) *Die Trajanssäule: Ein römisches Kunstwerk zu Beginn der Spätantike* (Berlin).

Lembke, K. (1994) *Das Iseum Campense in Rom. Studie über den Isiskulte unter Domitian* (Heidelberg).

Lepper, F. and Frere, S. (1988) *Trajan's Column. A New Edition of the Cichorius Plates* (Gloucester).

Lesquier, J. (1918) *L'armée romaine de l'Egypte d'Auguste à Dioclétien* (Cairo).

Levi, A., Caló and Levi, M. (1967) *Itineraria Picta. Contributo allo studio della Tabula Peutingeriana* (Studi e materiali del Museo dell'Impero romano 7) (Rome).

Levi, D. (1947) *Antioch Mosaic Pavements* (Princeton).

Levi, P. (1971) *Pausanias, Guide to Greece*, 2 vols (Harmondsworth).

Lewis, N. (1983) *Life in Egypt under Roman Rule* (Oxford).

Leylck, H. (1993) 'La vignetta di Antiochia e la datazione della Tabula Peutingeriana', *JAT* 3: 203–6.

Lieb, H. (1974) 'Zur Herkunft der Tabula Peutingeriana', in *Die Abtei Reichenau* (Neue Beiträge zur Geschichte und Kultur des Inselklosters) (Sigmaringen), 31–4.

Lloyd, A.B. (1975) *Herodotus Book II: Introduction* (Leiden).

—— (1976) *Herodotus, Book II: Commentary* 1–98 (Leiden).

—— (1993) *Herodotus, Book II: Commentary* 99–182 (Leiden).

Lomas, K. (1995) 'Urban elites and cultural definition: romanization in southern Italy', in T. Cornell and K. Lomas (eds) *Urban Society in Roman Italy* (London), 107–20.

—— (1996) *Roman Italy, 338 BC–AD 200. A Sourcebook* (London).
—— (1998) 'Roman imperialism and the city in Italy', in R. Laurence and J. Berry (eds) *Cultural Identity in the Roman Empire* (London), 64–78.
Lommatzsch, E. (1918) *Corpus Inscriptionum Latinarum* vol. I^2. *Inscriptiones Latinae Antiquissimae*, pars 2, fasc. I (Berlin).
Lopes Pegna M. (1952–3) 'Itinera Etruriae II', *SE* 22: 386.
Lowe, E.A. (1966) *Codices Latini Antiquiores: A Palaeographical Guide to Latin Manuscripts prior to the Ninth Century* vol. XI. *Hungary, Luxembourg, Poland, Russia, Spain, Sweden, the United States and Yugoslavia* (Oxford).
Luff, R.M. (1982) *A Zooarchaeological Study of the Roman North-western Provinces*, BAR International Series 137 (Oxford).
Luzzatto, G.I. (1962) 'Nota minima sul cosidetto *elogium* di Polla', *Studi in onore di Emilio Betti* 3 (Milan), 375–98.
Lynn, J.A. (ed.) (1993) *Feeding Mars. Logistics in Western Warfare from the Middle Ages to the Present* (Boulder).
—— (1999) *The Wars of Louis XIV, 1667–1714* (London).
McGing, B. (1998) 'Bandits, real and imagined, in Greco-Roman Egypt', *BASP* 35: 159–83.
McIntyre, J. and Richmond, I.A. (1934) 'Tents of the Roman army and leather from Birdoswald', *Transactions of the Cumberland and Westmoreland Antiquarian and Archaeological Society* 34: 62–90.
Mackay, C.S. (1999) 'Expressions to indicate goals of motion in the colloquial Latin of the empire', *ZPE* 126: 229–39.
Makkay, J. (1995) 'The treasures of Decebalus', *Oxford Journal of Archaeology* 14: 333–43.
Mann, J.C. (1987) 'A note on the so-called town-zone', *Britannia* 18: 285–6.
Mari, Z. (1996) 'Miliarium Aureum', in E.M. Steinby (ed.) *Lexicon Topographicum Urbis Romae* III (Rome), 250–1.
Marsden, E.W. (1969) *Greek and Roman Artillery. Historical Development* (Oxford).
—— (1971) *Greek and Roman Artillery. Technical Treatises* (Oxford).
Marshall, E. (1998) 'Constructing the Self and the Other in Cyrenaica', in R. Laurence and J. Berry (eds) *Cultural Identity in the Roman Empire* (London), 49–63.
Marshall, R. (1993) *Storm from the East. From Genghis Khan to Khublai Khan* (London).
Mattern, S.P. (1999) *Rome and the Enemy. Imperial Strategy in the Principate* (Berkeley).
Mattingly, D.J. (ed.) (1997) Dialogues in Roman Imperialism, *Journal of Roman Archaeology* Supplementary Series 23 (Portsmouth, RI).
Maxfield, V. (1996) 'The Eastern Desert forts and the army in Egypt during the principate', in D. Bailey (ed.) Archaeological Research in Roman Egypt, *Journal of Roman Archaeology* Supplementary Series 19 (Ann Arbor).
Maxfield, V.A. and Dobson, M.J. (eds) (1991) *Roman Frontier Studies 1989* (Exeter).
Meiggs, R. (1973) *Roman Ostia* (Oxford).
Mercuri, L. (1998) 'Les Eubéens et la Colonne de Rhègion', *MEFRA* 110: 551–69.
Meyboom, P.G.P. (1995) *The Nile Mosaic of Palestrina. Early Evidence of Egyptian Religion in Italy* (Leiden).
Meyer-Termeer, A.J.M. (1978) *Die Haftung der Schiffer im greichischen und römischen Recht* (Zutphen).
Mielczarek, M. (1993) *Cataphracti and Clibanarii. Studies on the Heavy Armoured Cavalry of the Ancient World* (Łódź).

Migne, J.-P. (1879) *Patrologia Latina* vol. 129 (Paris).
Millar, F. (1982) 'Emperors, frontiers and foreign relations, 31 BC–AD 378', *Britannia* 13: 1–23.
Miller, K. (ed.) (1887) *Die Peutingersche Tafel* (Ravensburg) (repr. Brockhaus, Stuttgart, 1962).
—— (1888) *Die Weltkarte des Castorius genannt die Peutingersche Tafel* (Ravensburg).
—— (1916) *Itineraria Romana. Römische Reisewege an der Hand der Tabula Peutingeriana dargestellt* (Stuttgart).
Millett, M. (1990) *The Romanisation of Britain* (London).
Milne, J.G. (1924) *A History of Egypt under Roman Rule* (London).
Milner, N.P. (1993) *Vegetius: Epitome of Military Science*, Translated Texts for Historians 16 (Liverpool).
—— (1998) 'A Roman Bridge at Oenoanda', *AS* 48: 117–23.
Mitchell, S. (1976) 'Requisitioned transport in the Roman empire: a new inscription from Pisidia', *Journal of Roman Studies* 66: 106–31.
—— (1986) 'Galatia under Tiberius', Chiron 16: 26–33.
Molin, M.M. (1990) 'Le transport du raisin ou du vin par la route à l'époque romaine en Gaule et dans les provinces voisines', in *Vigne*, 205–17.
Mommsen, T. (1851) 'Ueber die Unteritalien betreffenden Abschnitte der Ravennatischen Kosmographie', *BSGW, Philol.-hist. Classe* 3: 96–103.
—— (1856) 'Epigraphisches Miscellen: An Professor Ritschl', *RhM*. N.F. 10: 141–8
—— (ed.) (1883) *Corpus Inscriptionum Latinarum* vol. X. *Inscriptiones Bruttiorum, Lucaniae, Campaniae*, 2 vols (Berlin).
—— (1887) *Römisches Staatsrecht* II 2, 3rd edn (Leipzig).
Monte, M. del, Ausset, P. and Lefevre, R.A. (1998) 'Traces of ancient colours on Trajan's Column', *Archaeometry* 40: 403–12.
Morel, J.M.A.W. (1991) 'Tents or barracks?', in Maxfield and Dobson, 376–86.
Moreland, C. and D. Bannister, D. (1989) *Antique Maps*, 3rd edn (Oxford).
Morrison, J.S. and Coates, J.F. (1986) *The Athenian Trireme. The History and Reconstruction of an Ancient Greek Warship* (Cambridge).
Moscadi, A. (1970) 'Le lettere dell'archivio di Teofane', *Aegyptus* 50: 88–154.
Müller, C. (ed.) (1855–61) *Geographi Graeci Minores*, 2 vols (Paris).
Murphey, R. (1999) *Ottoman Warfare, 1500–1700* (London).
Mynors, R.A.B. (ed.) (1964) *XII Panegyrici Latini* (Oxford).
Nachtergael, G. (1988) 'Un aspect de l'environnement en Égypte gréco-romaine: les dangers de la circulation', *Ludus Magistralis* 21 (no. 64): 19–54.
Nicolet, C. (1991) *Space, Geography and Politics in the Early Roman Empire* (Jerome Lectures, 19) (Ann Arbor).
O'Connor, C. (1993) *Roman Bridges* (Cambridge).
Ohler, N. (1989) *The Medieval Traveller* (Woodbridge).
Olshausen, E. (1991) *Einfuehrung in die historische Geographie* (Darmstadt).
Oxé, A. (1926) 'Die römische Meile, eine griechische Schöpfung', *BJ* 131: 213–44.
Packer, J.E. (1997) *The Forum of Trajan in Rome. A Study of the Monuments* (Berkeley).
Pallu de Lessert, C. (1904–5) 'La syntaxe des routiers romains et les déformations des noms de lieux dans l'Afrique romaine', *MSNAF* 7e série 5: 115–38.
Palme, B. (1994) 'Flavius Sarapodorus, ein agens in rebus aus Hermopolis', *APF* 40: 43–68.

Panebianco, V. (1963–4) 'Il *lapis Pollae* e la partizione di *ager publicus* nel II sec. a.C., nel territorio dell'antica Lucania', *Rassegna Storica Salernitana* 24–5: 3–22.
Parker, G. (1972) *The Army of Flanders and the Spanish Road, 1567–1659. The Logistics of Spanish Victory and Defeat in the Low Countries' Wars* (Cambridge).
—— (1977) *The Dutch Revolt* (Harmondsworth).
—— (1988) *The Military Revolution* (Cambridge).
Patsch, C. (1937) *Der Kampf um den Donauraum unter Domitian und Trajan* (Vienna).
Peacock, D.P.S. and Maxfield, V.A. (1997) *Mons Claudianus. Survey and Excavation* (Cairo).
Peddie, J. (1994) *The Roman War Machine* (Stroud).
Pedley, T.G. and Winterbottom, S. (1991) *The Wooden, Leather and Bone Objects from Castle Street, Carlisle: Excavations 1981–82* (Carlisle).
Pekáry, T. (1968) *Untersuchungen zu den römischen Reichstraßen* (Antiquitas, Reihe 1. Abhandlungen zur alten Geschichte, Bd. 17) (Bonn).
Perjés, G. (1970) 'Army provisioning, logistics and strategy in the second half of the seventeenth century', *Acta Historica Academiae Scientarium Hungaricae* 16: 1–51.
Petersen, E., Domaszewski, A. von and Calderini, G. (1896) *Die Marcus-Säule auf der Piazza Colonna in Rom* (Monaco).
Petschenig, M. (ed.) (1913) *Sancti Ambrosii Opera, pars V. Expositio psalmi CXVIII* (Corpus Scriptorum Ecclesiasticorum Latinorum 62) (Leipzig).
Pflaum, H.G. (1940) *Essai sur le* Cursus publicus *sous le Haut-Empire romain*, (Mémoires présentés par divers savants à l'Académie des Inscriptions et Belles-Lettres de l'Institut de France, 14.1) (Paris).
Piccirillo, M. (1989) *Madaba: le chiese e i mosaici. Cinisello Balsamo* (Milan).
Pinkerton, J. (1808–24) *A General Collection of the Best and Most Interesting Travels in All Parts of the World*, 17 vols (London).
Pitimada, N. (1953) 'Rinvenimento di un cippo miliare', *NSA* serie 8a VII: 343–4.
Pitts, L.F. and St Joseph, J.K. (1985) *Inchtuthil. The Roman Legionary Fortress*, Britannia Monograph Series 6 (London).
Polaschek, E. (1965) 'Ptolemaios als Geograph', in *Pauly-Wissowa* Supplement X, 680–833.
Purcell, N. (1983) 'The apparitores: a study in social mobility', *PBSR* 51: 125–73.
Radke, G. (1973) 'Viae Publicae Romanae', *Real-Encyclopädie der Altertumswissenschaft*, Supplement XIII (A. Druckenmüller, Munich), cols 1415–686.
Raepsaet, G. (1999) 'Landtransport', in *Der Neue Pauly* VI, 1097–106.
Ramsay, A.M. (1920) 'Roman postal service under the republic', *JRS* 10: 79–86.
Rankov, B. (1990) 'Frumentarii, the Castra Peregrina and the provincial officia', ZPE 80: 176–82.
Rea, J.R., Salomons, R.P. and Worp, K.A. (1985) 'A ration-warrant for a adiutor memoriae', *YCS* 28: 101–13.
Rebuffat, R. (1986) 'Le Bouclier de Doura', *Syria* 63: 85–105.
—— (1995) 'A la recherche des milliaries perdus', *Mélanges Raymond Chevallier* vol. 2, *Histoire et archéologie* tome 2 [= *Caesarodunum* 29], Centre de Recherches A. Piganiol (Tours) 129–35.
Reddé, M. (1986) *Mare Nostrum. Les infrastructures, le dispositif et l'histoire de la marine militaire sous l'empire romaine* (Rome).
Reed, N. (1978) 'Pattern and purpose in the Antonine Itinerary', *AJPh* 99: 228–54.
Rees, B.R. (1968) 'Theophanes of Hermopolis Magna', *BRL* 51: 164–83.

Reincke, G. (1935) 'Nachrichtenwesen', *Real-Encyclopädie der Altertumswissenschaft*, vol. XVI.2 (Stuttgart), cols 1496–541.
Reise, A. (ed.) (1878) *Geographi Latini minores* (Heilbronn) (repr. Hildesheim, 1964).
Richmond, I.A. (1935) 'Trajan's army on Trajan's Column', *PBSR* 13: 1–40.
—— (1968) *Hod Hill, II, Excavations Carried out between 1951 and 1958* (London).
Rivet, A.L.F. and Smith, C. (1979) *The Place-names of Roman Britain* (London).
Robinson, H.R. (1975) *The Armour of Imperial Rome* (London).
Rockwell, P. (1981–3) 'Preliminary study of the carving techniques on the Column of Trajan', *Marmi Antichi*, Studi Miscellanei 26: 101–11.
Romaios, K. (1904–5) 'Laconia IV: the Hermai on the NE frontier', *BSA* 11: 137–8.
Rostovtzeff, M. (1946) 'Numidian horsemen on Canosa vases', *American Journal of Archaeology* 50: 263–7.
—— (1957) *The Social and Economic History of the Roman Empire* (Oxford).
Roth, J.P. (1999) *The Logistics of the Roman Army at War (264 BC–AD 235)* (Leiden).
Ruoff-Väänänen, E. (1975) 'The roads leading to Rhegium', *Arctos* 9: 93–8.
Saatmann, K., Jüngst, E. and Thielscher, P. (1939) *Caesars Rheinbrücke* (Berlin).
Şahin, S. (1994) 'Ein Vorbericht ueber den Stadiasmus Provinciae Lyciae in Patara', *Lykia – Anadolu akdeniz arkeolojisi* 1: 130–7.
Salway, P. (1981) *Roman Britain* (Oxford).
Sandys, J.E. (1919) *Latin Epigraphy. An Introduction to the Study of Latin Inscriptions* (Cambridge).
Sanzi di Mino, M.R. (ed.) (1998) *La villa della Farnesina in Palazzo Massimo alle Terme* (Rome).
Sarnowski, T. and Trynkowski, J. (1983) 'Legio I Italica–Liburna–Danuvius', in D. Planck (ed.) *Studien zu den Militärgrenzen Roms* III (Aalen), 536–41.
Scarborough, J. (1985) 'Criton, physician to Trajan: historian and pharmacist', in J.W. Eadie and J. Ober (eds) *The Craft of the Ancient Historian. Essays in Honour of Chester G. Starr* (Lanham), 387–405.
Schalles, H.G. (1988) 'Römische Anker: ein Neufund vom unteren Niederrhein', *Archäologie im Rheinland*, 88–9.
Scheiper, R. (1982) *Bildpropaganda der römischen Kaiserzeit unter besonderer Berücksichtung der Trajanssäule in Rom und korrespondierender Münzen* (Bonn).
Schillinger-Häfele, U. (1963) 'Beobachtungen zum Quellenproblem der Kosmographie von Ravenna', *BJ* 163: 238–51.
Schleiermacher, M. (1984) *Römische Reitergrabsteine. Die kaiserzeitlichen Reliefs der triumphierenden Reiters* (Bonn).
Schlüter, W. (1993) *Kalkriese – Römer im Osnabrücker Land. Archäologische Forschungen zur Varusschlacht* (Bramsche).
Schneider, K. (1935) 'Miliarium', *Real-Encyclopädie der Altertumswissenschaft*, Supplement VI (Stuttgart), cols 395–432.
Schnetz, J. (ed.) (1940) *Itineraria Romana* II: *Ravennatis Anonymi Cosmographia et Guidonis Geographica* (Leipzig) (repr. with index added by M. Zumschlinge, Stuttgart, 1990).
—— (1942) *Untersuchungen über die Quellen der Kosmographie des anonymen Geographen von Ravenna* (Sitzungsberichte der Akademie der Wissenschaften, Philosophisch-historische Abteilung 6) (Munich).
Schramm, E. (1980) *Die antiken Geschütze der Saalburg* (Bad Homberg).

Schroff, H. (1932) 'Tabellarii', *Real-Encyclopädie der Altertumswissenschaft*, vol. IV A.2 (Stuttgart), cols 1843–7.

Schweder, E. (1893) 'Ueber den Ursprung und die ältere Forme der Peutingerschen Tafel', *Jahrbuch für Philologie* 39: 485–512.

Sechi, M. (1990) *La costruzione della scienza geografica nei pensatori dell'antichità classica* (Memorie della Società Geografica Italiana 44) (Rome).

Settis, S. (ed.) (1988) *La Colonna Traiana* (Rome).

Sherk, R.K. (1974) 'Roman geographical exploration and military maps', *Aufstieg und Niedergang der römischen Welt* II, *Principat* 1 (Berlin), 534–62.

Sidebotham, S.E. (1986) *Roman Economic Policy in the Erythra Thalassa 30 BC–AD 217* (Leiden).

Sidebotham, S.E. and Zitterkopf, R.E. (1997) 'Survey of the via Hadriana by the University of Delaware: the 1996 season', *Newsletter of the American Research Center in Egypt*.

Skeel, C.A.J. (1901) *Travel in the First Century with Special Reference to Asia Minor* (Cambridge).

Smith, R.R.R. (1991) *Hellenistic Sculpture* (London).

Solin, H. (1981) *Zu lukanischen Inschriften* (Commentationes Humanarum Litterarum, 69) (Helsinki).

Southern, P. (1996) 'Men and mountains, or geographical determination and the conquest of Scotland', *Proceedings of the Society of Antiquaries of Scotland* 126: 371–86.

—— (1997) *Domitian. Tragic Tyrant* (London).

Speidel, M.P. (1975) 'The rise of ethnic units in the Roman imperial army', *Aufstieg und Niedergang der römischen Welt* II.3: 202–31.

—— (1985) 'Furlough in the Roman army', *YCS* 28: 283–93 = *Roman Army Studies* II: 33–41.

—— (1989) 'The soldiers' servants', *Ancient Society* 20: 239–47.

—— (1992) *Roman Army Studies* (Stuttgart).

—— (1994) *Riding for Caesar. The Roman Emperor's Horse Guard* (London).

Starr, C.G. (1941) *The Roman Imperial Navy, 31 BC–AD 324* (Chicago).

Stein, A. (1944) *Die Reichsbeamten von Dazien* (Budapest).

Steinby, M. (ed.) (1995) *Lexicon Topographicum Urbis Romae* II (Rome).

Stevens, C.E. (1937) 'Gildas and the Civitates of Britain', *English Historical Review* 52: 193–203.

—— (1951) 'The Roman Name of Ilchester', *Proceedings of the Somerset Archaeology and Natural History Society* 96: 188–92.

Stockton, D.L. (1981) *From the Gracchi to Sulla: Sources for Roman History, 133–80 BC* (LACTOR 13) (London).

Stoffel, P. (1994) *Uber die Staatspost, die Ochsengespanne und die requirierten Ochsengespanne* (Bern).

Stoiculescu, C.D. (1985) 'Trajan's Column's documentary value from a forestry viewpoint', *Dacia* 29: 81–98.

Stolte, B.H. (1949) *De Cosmographie van der Anonymus Ravennas: een studie over de bronnen van boek II–IV* (Amsterdam).

Strassi, S. (1997) *Le funzioni degli* [ἐπηρεται] *nell'Egitto greco romano* (Heidelberg).

Strauss, W.L. (1975) *The German Single-Leaf Woodcut, 1550–1600* (New York).

Strobel, K. (1984) *Untersuchungen zu den Dakerkriegen Trajans. Studien zur Geschichte des mittleren und unteren Donauraumes in der Hohen Kaiserzeit* (Bonn).

Stuart Jones, H. (1910) 'The historical interpretation of the reliefs of Trajan's Column', *Papers of the British School at Rome* 5: 435–59.
Stucchi, S. (1960) *Contributo alla conoscenza della Topographia dell'Arte e della Storia nella Colonna traiana, il viaggio marittimo di Traiano all'inizio della seconda guerra Dacica* (Udine).
Stueckelberger, A. (1994) *Bild und Wort: Das illustrierte Fachbuch in der antiken Naturwissenschaft, Medizin und Technik.* Kulturgeschichte der antiken Welt 62 (Mainz).
Susini, G., (1984) 'Le lapis de Polla', *Analecta Academiae voor Wetenschappen, Letteren en Schone Kunsten van België* 46: 110.
—— (1992) 'Per una classificazione delle iscrizioni itinerarie', in L. Quilici and S. Quilici Gigli (eds) *Tecnica stradale romana* (Università di Bologna Istituto di Archeologia. Cattedra di topografia dell'Italia antica; Atlante tematico di Topografia antica, 1) (Rome), 119–21.
Syme, R. (1988) 'Military geography at Rome', *Classical Antiquity* 7: 227–51.
Tarpin, M. (1999) '*Oppida vi capta, vici incensa* . . . les mots Latins de la ville', *Latomus* 58: 279–97.
Tchernia, A. (1986) *Le vin de l'Italie romaine* (Rome).
Thévenot, E. (1969) *Les voies romaines de la cité des Éduens* (Collection Latomus 98) (Brussels).
Thomas, J.D. (1976) 'The date of the revolt of L. Domitius Domitianus', *ZPE* 22: 253–86.
—— (2000) 'Communication between the prefect of Egypt, the procurators and the *nome* officials', in W. Eck (ed.) *Lokale Autonomie und römischen Ordnungsmacht in den kaiserzeitlichen Provinzen vom 1. Bis 3. Jahrhundert* (Oldenbourg), 181–95.
Thomasson, B.E. (1984) *Laterculi Praesidum I* (Gothenburg).
Thomson, J.O. (1948) *History of Ancient Geography* (Cambridge).
Tibiletti, G. (1955) 'Lo sviluppo del latifondo in Italia dall'epoca graccana al principio dell'Impero', *X Congresso internazionale delle scienze storiche, Roma, 1955* (Biblioteca storica Sansoni, n.s. 23) [= Xe Congrès international des sciences historiques, Rome, 1955 (Études présentées à la Commission internationale pour l'histoire des assemblées d'états, 18), Publications universitaires de Louvain, Louvain, 1958], *Relazioni* vol. II (Florence), 235–92.
Tomei, M.A. (1997) *Museo Palatino* (Milan).
Toynbee, J.M.C. (1973) *Animals in Roman Life and Art* (London).
Tozer, H.F. (1897) *History of Ancient Geography* (Cambridge).
Viereck, H.D.L. (1996) *Die römische Flotte. Classis Romana* (Hamburg).
Vigne (1990) *Archéologie de la Vigne et du Vin*, Caesarodunum 24 (Paris).
Villard, M.P. (1990) 'Les barbares et la boisson', in *Vigne*, 246–52.
Vine, B.H. (1993) *Studies in Archaic Latin Inscriptions* (Innsbrucker Beiträger zur Sprachwissenschaft 75) (Innsbruck).
Vulpe, R. (1963) 'Les bures alliés de Décébale dans la première guerre dacique de Trajan', *Studii Clasice* 5: 223–47.
—— (1969) 'Dion Cassius et la campagne de Trajan en Mésie Inférieure', *Studii Clasice* 6: 205–32.
Wacher, J. (1974/1995) *The Towns of Roman Britain* (London).
Waltzing, J.P. (1895) *Etude sur les corporations professionelles chez les Romains, depuis les origines jusq' a la chute de l'Empire d'occident* (Brussels).

Ward, G.C. (1991) *The Civil War. An Illustrated History* (New York).

Warmington, E.H. (1940) *Remains of Old Latin IV. Archaic Inscriptions* (Loeb Classical Library) (Harvard).

Wartena, J.R. (1925) *Inleidung op een uitgave der Tabula Peutingeriana* (Amsterdam), 46–56.

Watson, G.R. (1969) *The Roman Soldier* (London).

Weaver, P.R.C. (1972) *Familia Caesaris* (Oxford).

Weber, E. (1976) *Tabula Peutingeriana: Codex Vindobonensis 324. Kommentar / Vollständige Faksimil-Ausgabe im Originalformat*, Akademische Druck und Verlagsanstalt (Graz).

—— (1989) 'Zur Datierung der Tabula Peutingeriana', in H.E. Herzig and R. Frei-Stolba (eds) *Labor omnibus unus: Gerold Walser zum 70. Geburtstag dargebracht von Freunden, Kollegen und Schülern* (Historia Einzelschrift 60) (Stuttgart), 113–17.

Weigall, A.E.P. (1909) *Travels in the Upper Egyptian Deserts* (Edinburgh).

Weingarten, S. (1999) 'Was the Pilgrim from Bordeaux a woman? A reply to Laurie Douglass', *JECS* 7: 291–7.

Wessely, P. (ed.) (1735) *Vetera Romanorum Itineraria, sive Antonini Augusti Itinerarium* (Amsterdam).

White, K.D. (1984) *Greek and Roman Technology* (London).

Whittaker, C.R. (1994) *Frontiers of the Roman Empire. A Social and Economic Study* (Baltimore).

Wilkes, J.J. (1983) 'Romans, Dacians and Sarmatians in the first and early second centuries', in B.R. Hartley and J.S. Wacher (eds) *Rome and Her Northern Provinces* (Gloucester), 255–89.

Wilson Jones, M. (1993) 'One hundred feet and a spiral stair: the problem of designing Trajan's Column', *Journal of Roman Archaeology* 6: 23–38.

Wiseman, T.P. (1970) 'Roman republican road building', *PBSR* 38 [n.s. 25]: 122–53 [= *Roman Studies, Literary and Historical* (Collected Classical Papers 1), Francis Cairns, Liverpool, 1987, 126–57].

Woolf, A. (1998) 'Romancing the Celts: a segmentary approach to acculturation', in R. Laurence and J. Berry (eds) *Cultural Identity in the Roman Empire* (London), 11–24.

Woolf, G. (1998) *Becoming Roman. The Origins of Provincial Civilization in Gaul* (Cambridge).

Yavetz, Z. (1988) *Slaves and Slavery in Ancient Rome* (New Brunswick and Oxford).

Zahariade, M. and Bounegru, O. (1994) 'Roman ships on the Lower Danube (1st–6th centuries): types and functions', in C. Westerdahl (ed.) *Crossroads in Ancient Shipbuilding* (Oxford), 35–41.

Ziolkowski, A. (1995) '*Urbs direpta,* or how the Romans sacked cities', in J. Rich and G. Shipley (eds) *War and Society in the Roman World* (London), 69–91.

INDEX